J. Schulze   M. Homann

C₄-Hydrocarbons and Derivatives

Resources, Production, Marketing

Translation from the German by M. R. F. Ashworth

With 63 Figures and 89 Tables

Springer-Verlag Berlin Heidelberg New York
London Paris Tokyo Hong Kong

Prof. Dr.-Ing. Joachim Schulze

Technische Universität Berlin
Institut für Technische Chemie
FB Physikalische und Angewandte Chemie
Straße des 17. Juni 135, D-1000 Berlin 12

Dr.-Ing. Malte Homann

Deutsche Shell Chemie GmbH
Kölner Str. 6, D-6236 Eschborn

Translator:

Prof. Dr. M. R. F. Ashworth
Universität des Saarlandes
Fachrichtung 13.5, Organische und
Instrumentelle Analytik
D-6600 Saarbrücken

ISBN-13:978-3-642-73860-9   e-ISBN-13:978-3-642-73858-6
DOI: 10.1007/978-3-642-73858-6

Library of Congress Cataloging-in-Publication Data.
Schulze, Joachim.
C4-hydrocarbons and derivatives : resources, production, marketing / J. Schulze, M. Homann ; translation from the German by M. R. F. Ashworth.
Bibliography: p. Includes index.
ISBN-13:978-3-642-73860-9 (U. S.)
1. Hydrocarbons. I. Homann, M. (Malte), 1956 —. II. Title.
TP692.4.H9S33 1989 661'.81--dc20 89-6398 CIP

© Springer-Verlag Berlin Heidelberg 1989
Softcover reprint of the hardcover 1st edition 1989

Typesetting: Pustet, Regensburg

2151/3030-543210 — Printed on acid-free paper

Foreword

The book treats the C_4-hydrocarbons and their secondary products as a contribution to chemical engineering economics, applying this field of teaching and research to the technical processes for making and processing this group of products, so important to the chemical industry.

As early as the 1950s the then director of the Institute for Technical Chemistry of the Berlin Technical University, Professor Herbert Kölbel, took the initiative in the domain of Chemical Engineering Economics and began systematic studies of Project Engineering and Cost Estimation in connection with chemical plants. He also started a course on technical chemical processes in 1966. Properties, production procedures, plant equipment, and also the uses of technically interesting products are the central features of Chemical Technology. The information is to be found in the large encyclopedias of Technical Chemistry.

On the other hand, Chemical Engineering Economics deals with all the economic conditions of usage of the raw materials, possibilities of utilizing co-products, and the integration of these products into definite production programmes, from the standpoint of the chemical and technical fundamentals of the processes. Further important viewpoints are the costs of the products, taking into consideration important and variable influences on these costs, the situation and development of the market for the products and, of increasing significance, also the ecological global conditions for procuring raw materials and the production and marketing of the particular products.

Industrial chemistry is also touched on. This has been neglected in Western Europe and the German Federal Republic for a long time but some interesting books have been published in recent years. The present book provides, for the first time, a self-contained summary of industrial C_4-chemistry together with the peculiarities of Chemical Engineering Economics. The analysis of competitiveness of the processes and products in the C_4-field has been especially stressed.

We would like to express our thanks for the kind cooperation of Hüls AG, Marl, the leading company for industrial C_4-chemistry in Western Europe.

We would also like to thank Dipl. Ing. Werner Dabelstein of the Deutsche Shell AG, Hamburg, for his expert assistance in the chapter on fuel components; Dr. Wolfgang Jäkel of the Deutsche Shell Chemie GmbH, Eschborn, for his competent supplementary information concerning markets; and also Dipl. Chem. Peter Enders for the good cooperation with Springer-Verlag and for arranging for the translation into English.

We are further indebted to Hüls AG, Marl; ROW, Wesseling; and Deutsche Shell AG, Hamburg, for contributing several photographs.

Berlin and Eschborn, March 1989   J. Schulze · M. Homann

Contents

Introduction

In contrast to manufacturers who apply mechanical processing and finishing methods, the producers of chemical compounds can always markedly improve their competitive position when they succeed in making high-grade secondary products from the compounds for which there has been hitherto no, or only low-value, uses. Their special attention is thus directed towards finding a more profitable use — as economically and ecologically reasonable as possible — of the co-products which are unavoidably yielded.

The usual bottleneck on the way to solving this problem is in the development of suitable processes for utilizing these chemicals. It is not enough for the processes to be technologically suitable for profiting from these alternative starting materials; proof of their superior competitive ability is finally decisive for their usefulness.

Studies of the competitive power, especially intermediate- and long-term, of new chemical technologies or products are complicated and have thus been treated up to now in an inadequately scientific way. They consist of analyses and forecasts of the economic values of competing processes and products, and also take into consideration criteria of competition which cannot be evaluated by economic parameters. Evidence of economic success requires special consideration of marginal cost problems which are complicated when co-products arise; this is because co-products are yielded in different proportions and also usually of different type in differing processes and/or with other starting materials. Further, changes in processes and/or the choice of a different base chemical generally alter the property range of the products and hence their market value; the estimation of the benefits from the products to be prepared must be considered in relation to the evidence for economic success. The analyses of application, profitability and substitution necessary for this extend from the feedstocks right up to the end uses in the consumer goods sphere.

Competitive factors which cannot be evaluated economically also have an influence above all as ecological interests and political dependence on the procurement, production and/or sales chances of chemical products and the profitability of chemical engineering innovations.

Such analyses and forecasts therefore demand comprehensive, systematic knowledge along the lines of technology and ecology, production, strategies of procurement and utilization, cost calculation and political economy. It is true that in management science it is now stressed so that planning for success presupposes thinking of decisions in several dimensions, as required in scenario analyses design. So far there has been a lack of scientific work which has dealt with these problems of multi-layered tasks of decision making in practice.

The aim of the present investigation is, taking the example of the chemistry of the

C_4-hydrocarbons, to illuminate the complex problems concerning decisions about investment in new technologies or products; and then to present methods of evaluation suitable for providing a reliable basis of information in order to derive plans involving tremendous financial risks.

C_4-chemistry is highly suitable for presenting these methods of evaluation because it has mainly become a developed domain of large scale chemistry only recently, in contrast to the development of C_2- and C_3-chemistry with the chemical utilization of ethylene and propene, which took place in the 1960s and 1970s. Interesting from management and economic points of view are the prospects of international competitive activity from large companies involved in C_4-chemistry. At present, W. European and Japanese companies hold the leading position, whereas American and, lately, Arabian companies are still mainly involved with C_2/C_3-chemistry.

It is true that the large scale utilization of C_4-products began as early as the 1940s, with the production of petrochemical butadiene. The R and D work on utilizing further C_4-compounds, above all butanes and butenes, the inadequate exploitation of which was repeatedly lamented, followed with increasing activity in the 1970s and 1980s. The transformation of these technological innovations into profitable ventures is going to be largely a thing of the immediate future.

The raw material basis for C_2- and C_3-chemistry is still to a considerable extent ethane, obtained as a co-product when natural gas, and to some extent, crude oil are produced; by cracking practically only ethylene and propene are yielded. On the other hand, C_4-chemistry is based on the cracking of naphtha and also heavier fractions such as gas oil, which are formed at an advanced processing stage of the crude oil and are thus considerably more expensive. Naphtha cracking yields a variety of co-products, the exploitation of which requires special abilities regarding processing techniques, product application, R and D, and also a highly developed infrastructure for the possibilities of utilizing these secondary products.

American petrochemistry is centred mainly on ethane, that of W. Europe and Japan is based almost entirely on utilization of naphtha. For the Arabian countries now, in the 1980s, commencing the production of secondary petrochemicals, ethane is also relevant on a raw material basis since it has largely been flared there up to now and has had virtually no value. Many ethylene and propene derivatives which have been successful in C_2- and C_3-chemistry, can now be made on the basis of C_4-products which are yielded during naphtha cracking; hard international substitution competition is thus in the offing.

C_2/C_3-chemistry has the advantage of more favourable raw material prices in this competition. The principal superiority of C_4-chemistry lies in the chance of the chemists to exploit the variety of the naphtha cracking products. In this politically influenced competition the opposing elements are the abundance of raw materials of the countries rich in oil and natural gas and the wealth of ideas and technological talent which is necessary for the exploitation of the many promising pathways for processing and upgrading high-value end products from C_4 raw materials.

The potentials for competition which arise from changing the production scheme through integration of new processes for C_4-chemistry are first systematically discussed here and followed by an assessment of their techno-economic influence on the competitive power of the relevant end products. There are regional differences, sometimes marked, in conditions of procurement, production and sales. Studies must thus also be made of the conclusions which different countries have to draw from the changes in the technological and marketing conditions.

This complex problem of the relationships of the competitive potentials is treated in five sections:

1) Presentation of the existing scheme of supply of C_4-hydrocarbons, with quantitative information about their availability worldwide and in the German Federal Republic (Chapter 2).
2) Description of the current situation regarding the processing technique for C_4-hydrocarbons, taking into account the present state of R and D work so far as this was published in the literature before the end of 1987. The relevance of literature information for practice has been checked in numerous discussions with experts in industry (Chapter 3).
3) Compilation of data concerning present worldwide processing streams of C_4-hydrocarbons and presentation of the consumption patterns (Chapter 4).
4) Management and politico-economical parameters influencing the economic success of new technical products (Chapter 5).
5) Evaluation of the competitive chances of C_4-hydrocarbon derivatives, demonstrated with the examples of the motor fuels and plastic market sectors (Chapter 6).

1 Fundamentals of C$_4$-Hydrocarbons

1.1 Nomenclature and Chemical Structure of the C$_4$-Hydrocarbons

Depending on the number of attached hydrogen atoms there are C$_4$-hydrocarbons which are saturated or contain double or triple bonds. Table 1.1 contains all the C$_4$-isomers, arranged in the order of increasing unsaturation.

Table 1.1. Nomenclature and structure of the C$_4$-hydrocarbons

Nomenclature	Abbreviated name	Empirical formula	Structural formula
Alkanes		C$_4$H$_{10}$	
Butane	*n*-Butane		CH$_3$–CH$_2$–CH$_2$–CH$_3$
Isobutane	*i*-Butane		(CH$_3$)$_2$CH–CH$_3$
Alkenes		C$_4$H$_8$	
1-Butene			CH$_2$=CH–CH$_2$–CH$_3$
cis-2-Butene	*n*-Butene		CH$_3$–CH=CH–CH$_3$
trans-2-Butene			CH$_3$–CH=CH–CH$_3$
2-Methylpropene	*i*-Butene		(CH$_3$)$_2$C=CH$_2$
Dienes		C$_4$H$_6$	
1,2-Butadiene	Butadiene		CH$_2$=C=CH–CH$_3$
1,3-Butadiene			CH$_2$=CH–CH=CH$_2$
Alkynes		C$_4$H$_6$	
1-Butyne	(Ethylacetylene) Butyne		CH≡C–CH$_2$–CH$_3$
2-Butyne	(Dimethylacetylene)		CH$_3$–C≡C–CH$_3$
Alkenynes		C$_4$H$_4$	
1-Butyne-3-ene	Vinylacetylene		CH≡C–CH=CH$_2$

1.2 Definition of the C$_4$-Cut from Steam Crackers and Its Raffinates

The C$_4$-cut from steam crackers is the term for a mixture which contains about 20 different C$_3$-C$_5$-hydrocarbons, the boiling points of which lie between $-47.7\,^{\circ}$C and $26.9\,^{\circ}$C under normal pressure (see Table 3.1). It is obtained along with ethylene, propene and further products, in the low temperature fractionation of liquid hydrocarbons from steam crackers. The yields of the individual products depend on the nature of the raw materials used and the cracking conditions (Table 1.2).

Table 1.2. Typical composition of the C$_4$-cut from steam crackers and its raffinates, using liquid cracking raw materials (in wt. %) [1.1]

	C$_4$-Cut	Raffinate I	Raffinate II
i-Butane	0− 2	0− 4	0− 7
n-Butane	2− 5	4−12	7−20
i-Butene	18−32	35−55	0.1
1-Butene	14−22	25−35	45−60
2-Butene	5−15	10−25	18−42
Butadiene	35−50	0.5	1

The 1,3-butadiene is usually first selectively separated from the C$_4$-cut. The residual product, free of butadiene, is termed raffinate I. Removal of the i-butene from this yields raffinate II, which consequently contains only butanes and n-butenes.

1.3 Terms for the C$_4$-Hydrocarbon Streams Within the Refinery

During the refining process of crude oil, gases are yielded which contain the butanes and butenes but scarcely any butadiene. Their composition differs from those of the C$_4$-cut and raffinates I and II, and the C$_4$-hydrocarbons encountered in refining processes are termed refinery B-B and refinery gases. Refinery B-B means a stream mostly from catalytic crackers, containing butenes and butanes; the butene concentration is well below that in the streams from steam crackers (Table 1.3). The composition of the refinery gases is subject to considerable fluctuations; this depends on the structure of the refinery processing plants.

Table 1.3. Typical composition of refinery B-B from catalytic crackers (in wt. %) [1.1]

i-Butane	35−45
n-Butane	7−12
i-Butene	10−20
1-Butene	9−12
2-Butene	20−29.5
Butadiene	0− 0.5

1.4 Product Specifications

The individual procedures for subsequent processing of the C_4-cut depend on the concentrations of the various components. In C_4-chemistry there are a few general quality requirements but many additional individual specifications for the different processes must be considered. Since the major part of the C_4-cut subsequently finds captive use, most quality requirements are only of internal works interest.

Only 1,3-butadiene, i-butene and 1-butene come onto the market as intermediates with standardized product purities. When for example butadiene is to be used as a monomer for preparing cis-1,4-polybutadiene, purities are demanded which limit the content of acetylenes and allenes to 30 ppm in each case. For synthesis of SB-rubber, however, acetylene amounts up to 1000 ppm are tolerated and there are no prescribed limits for allene content.

The permissible contents of impurities in the usual commercial intermediates have fallen continually. Steady improvements in analytical techniques have contributed decisively to this. Table 1.4 shows how the demands on butadiene for stereospecific polymerisation have risen during the last ten years.

It can be seen that the impurities have been more than halved. The small amounts mean that there are high technical requirements for the separation procedures. Table 1.5 shows the presently valid specifications for high purity i-butene and 1-butene.

Table 1.4. Specifications for butadiene in poly-cis quality in 1975 [1.2] and 1985 [1.3]; data as minimum wt. % or maximum ppm by wt

Component	Concentration	
	1975	1985
1,3-Butadiene	99.5%	99.6%
Propyne	30 ppm	5 ppm
Acetylenes	50 ppm	30 ppm
Allenes	50 ppm	30 ppm
Dimers	500 ppm	250 ppm
Carbonyls (as acetaldehyde)	50 ppm	25 ppm
Peroxides (as H_2O_2)	5 ppm	1 ppm
Sulfur (as H_2S)	2 ppm	1 ppm

Table 1.5. Compositions of high purity i-butene and 1-butene, Hüls-commercial quality, 1985; data in wt. % or ppm by wt [1.4]

Component	i-Butene	1-Butene
i-Butene	99.98%	0.15%
1-Butene	0.005%	99.70%
2-Butene	0.01%	0.01%
Butane	0.005%	0.15%
1,3-Butadiene	< 10 ppm	< 10 ppm
tert-Butanol	< 5 ppm	< 1 ppm
Water	< 30 ppm	< 10 ppm
Sulfur	< 1 ppm	< 1 ppm

Practically all the end products prepared from the C$_4$-cut are commercially available. General quality directives exist for them. For commodities (e.g. MTBE) they include mostly constant minimum requirements. For specialities there are, in addition to the standard features, usually additional criteria of quality adapted to the individual wishes of the customer.

2 Production Scheme for C$_4$-Hydrocarbons

2.1 Methods for Establishing the Production Scheme

The production scheme is intended to display the availability of amounts of raw materials, base chemicals or intermediates, together with their natural or synthetic sources and technological relations. It forms the basis for analyzing and forecasting the quantitative suitability and availability of a raw material for preparing chemical products.

The inquiry presupposes extensive knowledge of natural resources of raw materials and the possibilities of using them. Logistic viewpoints must be included in the cases of not easily accessible or remote areas. In order to establish the production scheme of base chemicals and intermediates, consideration must be given to all syntheses in which they are yielded as chief or co-products within chemical production or in production operations of other industries (refineries, coke oven plants), quoting production amounts or capacities and degrees of utilization.

Since pure resources are rarely involved the amounts of the particular starting chemicals must be derived using analytical data, empirical concentration values or technical yield coefficients. This should be supplemented by showing the preparation of important derivatives, even when they are not yet manufactured in practice. The degree of complexity is generally high and it is advisable to present the data as diagrams or tables in matrix form which indicate the amounts available.

Sources of information about production schemes and amounts are widely scattered in text books, encyclopedias, and specialist journals of Chemical Technology, Industrial Chemistry and Petrochemistry; in sales literature issued by mineral oil and chemical companies; in statistics of private associations and government departments (Table 4.1); and studies of marketing research institutes (Table 4.2). A problem arising from the continuously changing of market structures is that the data quickly cease to be relevant. Electronic data processing can be helpful in maintaining up-to-date information.

2.2 Production of C$_4$-Hydrocarbons

Crude oil and natural gas are the starting materials for preparing C$_4$-hydrocarbons. Figure 2.1 depicts their processing scheme.

In addition it is possible to obtain them by special synthetic routes in the fields of

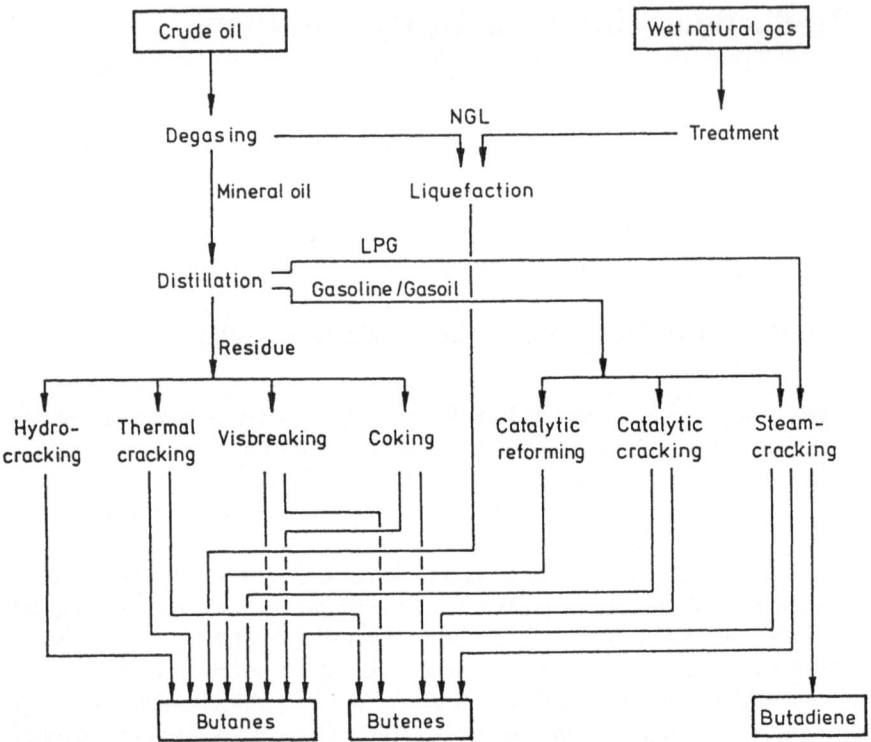

Fig. 2.1. Processing scheme for C₄-hydrocarbons from refinery and petrochemical processes

petrochemistry, coal chemistry and biochemistry; so far, however, these procedures have not yet gained commercial importance.

Wet natural gas is of primary interest as a source of the saturated C₄-hydrocarbons (C₄-paraffins) on account of its high propane and butane contents. Propane/butane mixtures (LPG) can be obtained during processing of the natural gas; the butanes are then derived by separation. Butane-containing gases are also yielded during crude oil production. They largely separate directly at the bore hole. Sometimes they are first recycled to maintain the oil reservoir pressure and only after the oil reserves have been exploited to a certain level are they collected or burnt because of restricted pipeline capacities. Small amounts are carried in solution with the crude oil as a transport medium (spiking). Additional amounts of butanes are yielded during atmospheric crude oil distillation and during some subsequent processes in refineries (reforming, cracking).

The unsaturated C₄-hydrocarbons (C₄-olefines) are more interesting for industrial uses than the saturated ones because of their more favourable properties. Up to now, they have been obtained almost entirely by cracking processes from two sources: they are yielded as co-products when the base products ethylene/propene are made in steam crackers (i. e. in the chemical industry); and, in contrast to other olefines, also during cracking processes of the refineries (i. e. in the oil industry). Steam crackers are considered, at present, to belong to the chemical industry; they represent a vertical backward integration of chemical companies into the refinery field. To this field belong the procedures of mineral oil processing and of refining to derive other products (lubricating oil, bitumen, etc.). In refinery processes C_4-olefine-containing cracked gases (refinery B-B) occur essentially in catalytic and thermal cracking installations. Processes of coal chemistry scarcely play a part in olefine production at present. In connection with the supply of raw materials they deserve interest in the context of studies about long-term development tendencies of raw material resources.

Table 2.1 shows the world production of C_4-monoolefines. It is evident that in North America by far the major part of the i-butene comes from refinery processes, whereas in Western Europe its origin is mainly as a steam cracking co-product. In contrast, n-butenes are derived in all cases from refinery processes. Table 2.9 contains data about production of butadiene.

Table 2.1. World production of butenes in 1984, in 1000 t/annum [2.1]

Area	i-Butene from				n-Butene from			
	Raff. I	Ref. B-B	Dehydr. Butane	Arco-TBA	Raff. II	Ref. B-B	Dehydr. Butane	α-Olefine Plant
N. America	590	4420	0*	250	540	9180	30	40
S. America	120	710	0	0	100	1590	60	0
W. Europe	920	730	0	50	830	1620	0	0
E. Europe	330	320	50	0	290	610	600	0
Remainder	580	920	0	0	540	2020	40	10
Total	2540	7100	50	300	2300	15020	730	50

* recently also in Texas as precursor of MTBE-preparation

2.2.1 C₄-Hydrocarbons from Natural Sources

C_4-Hydrocarbons (butanes) are directly available from natural sources. They are yielded in the petroleum oil fields as an associated gas because the oil has to be degased before transport. The associated gas contains, above all, methane, ethane, propane and butane; higher hydrocarbons remain in the oil as liquid components. A large amount of these gases is simply flared in the oil field areas. It was reported in 1981 that about 400 million m^3 per day of associated gas were burnt off in this way in the 18 largest oil-producing countries [2.2]. This corresponds in energy content to about 2.6 million barrels of crude oil per day or, according to the energy amount, about 8% of the oil production at that time. In the British off-shore sector of the North Sea in 1980, 7000 million m^3 of gas were burnt off because of limited pipeline capacity [2.3].

The second natural source consists of the so-called wet natural gas fields, among which the North Sea fields are particularly relevant for Northwest Europe on account of their geographical situation (Fig. 2.2). Wet gases have a different composition from the dry natural gas which predominates in nature. (see Table 2.2).

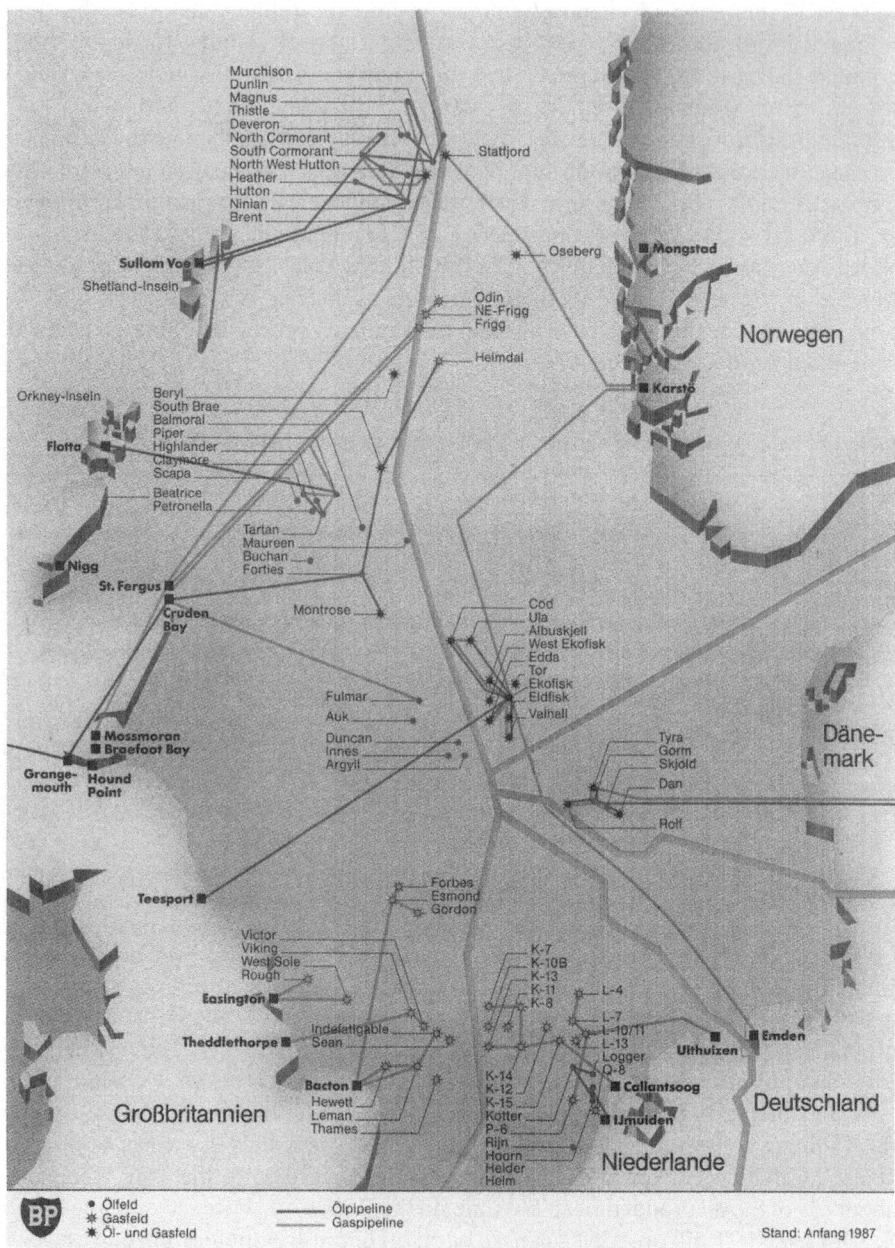

Fig. 2.2. Crude oil and natural gas pipelines from North Sea oil and gas fields [2.5]

Table 2.2. Composition of wet and dry sour and sweet gases (Vol. %) [2.4]

Type of Gas		CH$_4$	C$_2$H$_6$	C$_3$H$_8$	C$_4$H$_{10}$	C$_{5+}$	CO$_2$	H$_2$S	N$_2$
Sour	dry	81	1	–	–	–	8	6	4
	wet	54	19	7	3.5	1	9	6	0.5
Sweet	dry	80	2.7	0.8	0.3	0.1	6	0.001	10
	wet	62	12	11.5	4	0.5	10	–	–

Associated and wet gas together form the NGL[1]-pool, from which, through liquefaction by cooling, almost the entire LPG[2], plus 80% of the ethane dissolved in it, can be recovered in economically acceptable amounts.

Utilization of these resources demands either processing on the spot or a logistic system (collection-transport/storage-distribution); this was set up in Europe only a few years ago. The present low market prices for crude oil and rigid tax burdens on oil production are retarding the extension of important gas collecting sites in the North Sea.

NGL is delivered under pressure via a gas or oil pipeline to a separating installation where it is usually split into the fractions LPG and methane/ethane. The latter fraction is transported further by only relatively few established transport connections (e. g. by ship from the NGL-separating plant in Teesside, England to the Norethyl steam cracker in Bamble, Norway); LPG, however, has developed into a world-wide commercial product during the last few years as a result of the rapid expansion of special tanker fleets.

At present, the most important exporters of LPG are Saudi Arabia, Algeria and the North Sea countries i.e. the United Kingdom and Norway. By the end of the present decade Kuwait, the United Arabian Emirates, Canada, Mexico and Venezuela, as further potential exporters, will augment appreciably the LPG supply (Table 2.3).

Because of falling crude oil prices the planning data for LPG-production from North Sea-NGL have had to be reduced; various projects (e. g. the development of "Troll", the largest natural gas field in the North Sea) are profitable only when crude oil prices are high. Figure 2.3 gives a forecast of the future LPG-production in the North Sea according to the state of planning of 1981.

It is legally prescribed in the United Kingdom and Norway that for LPG from local sources priority must be given to the home chemical industry. Based on the now known figures for LPG-processing capacities at present and planned for the future, continued export surpluses from these countries can be expected. The export surpluses will consist almost entirely of butane from Norway, and practically only propane from the United Kingdom [2.3]. A syndicate managed by Ruhrgas is

---

[1] NGL = Natural Gas Liquids: liquefied gas mixture of ethane, propane, butane, condensate (C$_{5+}$-hydrocarbons)

[2] LPG = Liquefied Petroleum Gas: liquefied gases propane, butane or mixture

Table 2.3. Export disposition of LPG in 10^6 tons [2.6]

	1980	1985	1990
Saudi Arabia	6.7	12.8	12.9
Remaining Near East	5.7	14.2	13.6
Near East	12.4	27.0	26.5
United Kingdom	0.6	4.3	4.4
Norway	–	–	–
North Sea	0.6	4.3	4.4
Algeria	1.7	5.4	6.3
Libya	0.6	0.6	0.6
Nigeria	–	–	0.8
Africa	2.3	6.0	7.7
Australia	1.7	0.8	–
Indonesia	0.8	0.7	0.6
Far East	2.5	1.5	0.6
Canada	2.8	2.3	2.0
Mexico	0.3	2.5	1.0
Venezuela	1.2	1.5	1.0
North and South America	4.3	6.3	4.0
Others	–	0.5	2.0
World Total	22.1	45.6	45.2

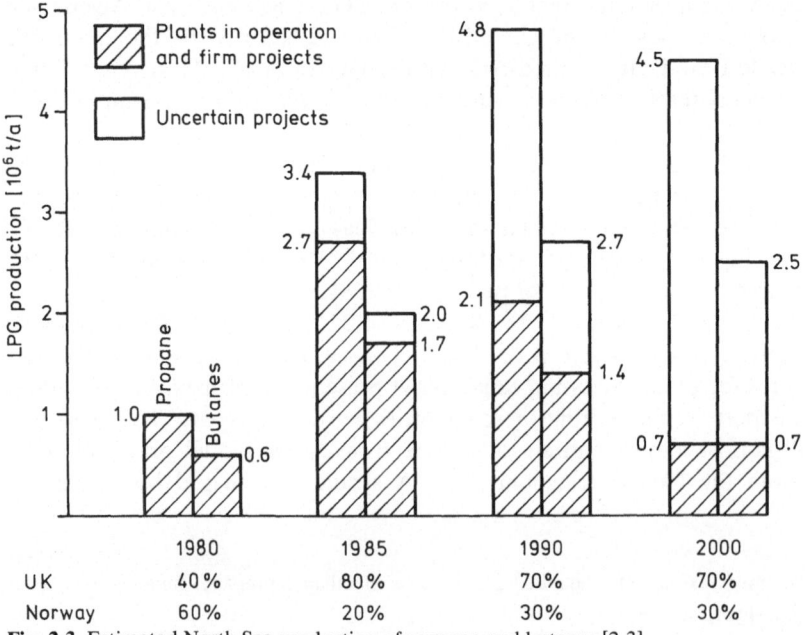

Fig. 2.3. Estimated North Sea production of propane and butanes [2.3]

operating a natural gas terminal (fractionation and storage) in Emden which has been connected, since 1975, with a natural gas pipeline (91 cm diameter) to the Ekofisk area 450 km away.

The sales chances for LPG are determined not only by the amounts offered but also, and mainly, by its price. The price depends on the costs of obtaining and transporting the LPG and decisively also on the fiscal burdens (customs and taxes) of the importing countries.

The olefine plants near the coast, requiring LPG (e. g. steam crackers of Dow in Terneuzen, of Shell in Pernis and Moerdijk, and of Petrochim in Antwerp) will for the present have a decisive influence on the market price in Western Europe. Thus in a calculated example, LPG as a base chemical was preferable to naphtha only when the price (cif) amounted to not more than 83% (propane basis) to 86% (butane basis) of that of naphtha (in U.S. dollars/ton). This was derived from the yield pattern from cracking and the evaluation of the steam cracker products at usual market prices [2.7]. A comparison of the starting materials LPG and crude oil can give an estimate of the competitive position of LPG for the preparation of ethylene.

In Table 2.4 the equivalent price ratios for propane and butane have been calculated, below which the preparation of ethylene from LPG is advantageous. The calculation is based on the average prices for naphtha and crude oil in 1983.

Figure 2.4 shows the price policy for the most important LPG and crude oil producers so far. The curve of the equivalent price ratio shows the distinct advantages in the competitive position of crude oil up to 1986. These were especially evident during the second oil crisis of 1979.

Table 2.4. Evaluation of the necessary LPG/crude oil-equivalent price ratio in the Arabian Gulf for ethylene production in Northwest Europe in 1983 [2.7]

Basis	Propane	Butane
Ø Naphtha contract price, $/t	277	
LPG/naphtha equivalent price ratio* (cif)	0.83	0.86
theor. LPG-price (cif), $/t	229	239
less LPG-cargo rate. Arab. Gulf-Rotterdam, $/t	35	35
theor. LPG-price (fob), $/t	194	204
Ø crude oil price (fob) "Arabian Light", $/bbl	29	
Ø crude oil price (fob) "Arabian Light", $/t	215	
LPG/crude oil-equivalent price ratio (fob)	0.90	0.95

* for ethylene production in existing flexible steam crackers
  cif = cost, insurance, freight
  fob = free on board

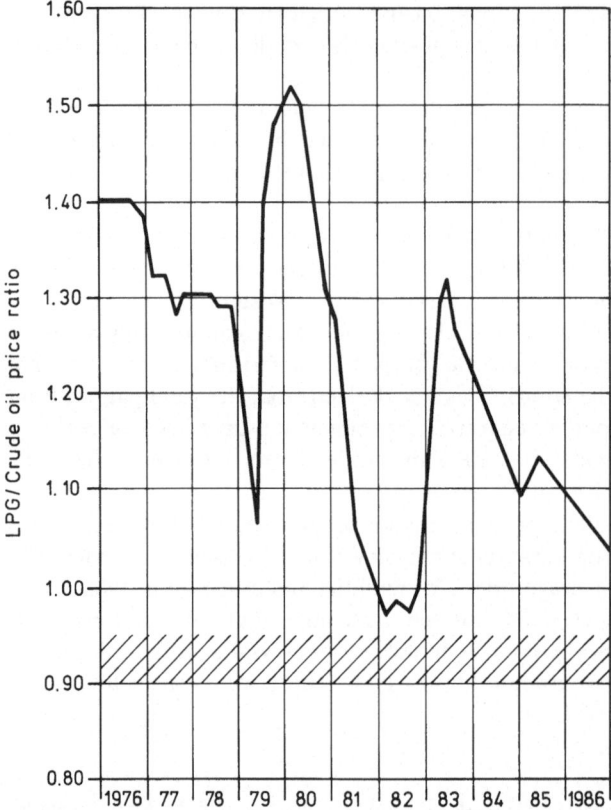

Fig. 2.4. LPG/crude oil price ratio (fob) in the Arabian Gulf, 1976–1986 [2.7; 2.8]

2.2.2 C$_4$-Hydrocarbons from Refinery Gases

Inside information about the qualities of the crude oil used and the corresponding basic yields is needed to be able to quantify the C$_4$-paraffins yielded during distillation of the crude oil. The sources, as an indication of the quality, amounts and prices of crude oils are published in the Federal German Republic in the journal Erdöl Informationsdienst (EID) [2.9]. According to an expert estimate, the refinery gas yields of top distillation are in the order of 2% [2.10].

The following average composition by volume of these refinery gases, obtained purely by distillation, can be given, without claim for general validity [2.11]

Refinery gas	Methane	Ethane	Propane	Butanes	Pentanes
Vol. %	5	10	30	35	20

In an analysis of Federal German refineries for yields of C$_4$-olefines one has first to ascertain the capacities for catalytic and thermal cracking processes. Data for these are compiled in Table 2.5.

Table 2.5. Annual capacities for processes yielding refinery B-B in the FRG, 1985 [2.12]

Refinery (company, site)	Crude oil cap. 10^6 t	Catalytic cracking 10^3 t	Thermal cracking 10^3 t	Coking 10^3 t	Visbreaking 10^3 t
BP, Vohburg	5,1	885			
ERN[1], Neustadt	7	1200			800
Esso, Hamburg	4,5	750			
Esso, Ingolstadt	5	1200			
Esso, Karlsruhe	7,5		1280	770	1570
Marathon, Burghausen	3,4			1400	
Mobil, Wörth	3,5	810			
OMW[2], Karlsruhe	7	2700			1500
Shell, Cologne-Godorf	9		1925		
Shell, Hamburg	4,5	620	820		
Texaco, Heide	4	430			
URBK[3], Wesseling	4,5		2000		
Veba, Gelsenk.-Buer and Horst	10,5	840		1360	970
Wintershall, Lingen	4,5			840	
Total	80,0	9435	6025	4370	4840

[1] 50% Veba-Oil (Aral), 50% Mobil Oil
[2] 42% German Texaco, 33% Veba-Oil (Aral), 25% Conoco (Jet)
[3] 75% Rheinbraun, 25% Hoechst

With four large refineries with catalytic and thermal crackers ceasing to operate during the last three years (BP/Hünxe, Esso/Cologne, Shell and ERI[1]/Ingolstadt) and, in addition, the BP refinery in Hamburg not operating at present, the distillation capacity of refineries with conversion equipment amounted, in 1985, to about 80 million tons crude oil throughput (Table 2.5). Adding to this the residual capacities of the last two hydroskimming refineries ERD[2], Duisburg (2 million t) and ERM[3], Mannheim (3.5 million t) gives for 1985 an average total capacity of crude oil throughput of about 85.5 million tons. A crude oil charge in the atmospheric distillation of 70 million tons is assumed for the purpose of estimating the degree of capacity utilization, which comes to 82%.

With this degree of utilization and the characteristic yield factors for C$_4$-base chemicals from the refinery processes, which are quoted in Table 2.6, the refinery C$_4$-amounts of Table 2.7 can be calculated. Added to this are the estimated amounts of butanes which are yielded during the top distillation and in the hydrocrackers of Shell, Godorf (990,000 t/a), URBK, Wesseling (900,000 t/a) and Wintershall, Lingen (1 million t/a) at medium cracking severity. From these production data one can see that the catalytic cracking plants have the biggest share within the total refinery B-B production. Figure 2.5 shows the catalytic cracker of the Shell refinery in Wilmington/USA.

[1] 50% Agip Germany, 50% Gelsenberg
[2] 80% German Fina, 20% German Total
[3] 100% Wintershall

Table 2.6. C$_4$-Yields in refinery processes [2.13]

Yield/process (wt. %)	Catalytic cracking	Thermal cracking	Coking	Visbreaking
Refinery B-B	9.0	7.5	2.0	1.0
of which *n*-butenes	2.9	1.8	0.4	0.3
i-butene	1.7	1.2	0.7	0.2
butanes (80% *i*-butane)	4.4	4.5	0.9	0.5

Table 2.7. C$_4$-Base chemicals from refinery processes, FRG, 1985 (in 1000 t/a), assuming 82% capacity utilization

Refinery gases			
from top distillation	1400	of which butanes	490
from hydrocrackers	310		200
refinery B-B			
from cat. crackers	696	of which *n*-butenes	339
therm. crackers	370	*i*-butene	225
coking	72	butanes	614
visbreaking	40		
Total refinery B-B	1178		1178
Total C$_4$-base chemicals from refineries			1868

Fig. 2.5. Catalytic cracker of the Shell refinery in Wilmington/USA

This model calculation for the German Federal Republic can easily be adapted to other countries with the aid of capacity data, degrees of utilization of the refinery installations and the generally applicable values of the yield factors.

2.2.3 C$_4$-Cut as Co-product of Ethylene Production

The amount of the C$_4$-cut falling into the chemical sector can be ascertained from the co-production of the C$_4$-cut related to ethylene production. The spectrum of the yield of co-products from steam cracking, of prime importance for the profitability of ethylene production, depends essentially on the raw materials and the cracking conditions (Table 2.8). The raw materials for steam cracking can be roughly classified into liquefied gases from natural gas fields, naphtha and gas oil fractions from refineries. Their choice is governed by regional considerations and in the following calculations for the purpose of investigating present day schemes are taken as given.

The amount of C$_4$-cut provided from ethylene crackers in the Federal Republic of Germany in 1985 has been calculated in Table 2.9. Based on the steam cracker capacities the ethylene capacity attains a total of about 4 million tons. According to official information (Stat. Bundesamt, FRG), the ethylene production in 1985 was 3 million tons, corresponding to a utilization factor of approx. 75% for the crackers. In order to attain the effective capacity from the nameplate capacity about 200,000 t/a had to be subtracted because withdrawal from operation of a ROW cracker in 1985; the real cracker utilization is thus higher. No account has been taken of the ethylene resulting as co-product from acetylene production using the electric arc process at Hüls and using the Sachsse process at BASF.

The raw material for cracking in 1985 in the German Federal Republic consisted of about 75% naphtha, plus a remainder from refinery co-products with, to some extent, similar cracking properties. Bearing in mind the present energy price level and the increased demand for propene of recent years, medium severity cracking proved best for the naphtha. Using the yield characteristic factors of Table 2.8 and the assumptions quoted, the quantities of the C$_4$-cut and its components have been calculated for 1985 and are given in Table 2.9.

Figure 2.6 illustrates a steam cracking plant (ROW, Wesseling, steam cracker No. 4).

Table 2.8. C$_4$-yields during steam cracking of various feedstocks without ethane recycle [2.13]

Yield/raw materials (wt. %)	Ethane	Propane	n-Butane	Naphtha			Atm. gas oil	Vacuum gas oil
				ls	ms	hs		
Ethylene	80	32	30	22	27	30	23	19
C$_4$-cut	1.1	4.5	12.3	12.7	11.5	8.2	8.6	7.9
of which butadiene	very	1.9	3.5	4.8	5.0	3.2	3.9	3.6
n-butenes	small	1.1	1.8	3.5	2.9	2.1	3.1	2.8
i-butene	amounts	1.4	2.0	4.0	3.2	2.3	1.5	1.4
butanes		0.1	5.0	0.4	0.4	0.3	0.1	0.1

Cracking conditions: ls = low severity; ms = medium severity; hs = high severity

Table 2.9. Steam cracker capacities, raw materials and C_4-hydrocarbons obtained from ethylene production in the FRG in 1985 (in 1000 t/a). Cracker capacity utilization 75%; medium severity naphtha cracking conditions

Firm, site	Ethylene capacity	Raw material	C_4-Products				
			C_4-cut	butadiene	n-butenes	i-butene	butanes
BASF, Ludwigshafen	450	Naphtha	144	63	36	40	5
Caltex, Raunheim[1]	250	Naphtha	80	35	20	22	3
EC, Cologne-Worringen	720	Naphtha	230	100	58	64	8
Esso Chemie, Cologne[1]	450	Naphtha	144	63	36	40	5
Marathon, Burghausen	300	Naphtha	96	42	24	27	3
ROW, Wesseling	650	Waxy dist.[2]	143	62	36	40	5
Texaco, Heide	80	Naphtha	25	11	6	7	1
URBK, Wesseling[3]	520	atm. gas oil	146	66	53	25	2
Veba, Gels.-Buer	400	Naphtha	128	56	32	36	4
Veba, Münchsmünster	200	Naphtha	64	28	16	18	2
Total	4020		1200	526	317	319	38

[1] The steam crackers of Caltex and Esso Chemie were shut down as part of the adaptation to the demand for ethylene
[2] Wax distillate from hydrogenated vacuum gas oil of Shell, Godorf, with cracking yields as medium severity cracked naphtha
[3] Hydropyrolysis, sometimes naphtha also as raw material

Fig. 2.6. Steam cracker of ROW, Wesseling

Table 2.10. Butadiene extraction capacities in the FRG, 1985

Company, Site	Extraction agent	Extraction capacity in 1000 t butadiene
BWH, Marl	NMP	120
EC, Cologne-Worringen	NMP	245
Esso Chemie, Cologne	CAA	23
ROW, Wesseling	NMP	130
Total		518

NMP = N-methylpyrrolidone, CAA = cupric ammonium acetate

Butadiene has been extracted from this C$_4$-cut at four central locations in the Federal Republic (Table 2.10). Especially BWH, EC and ROW are supplied with C$_4$-cut for butadiene extraction purposes from several ethylene producers, via long-distance pipeline or tank vehicles. Extraction capacity here is rather low with reference to the butadiene contained in the C$_4$-cut and the question arises whether those ethylene producers who are at a disadvantage regarding transport costs, should make use of this extraction process. In any case it can be concluded that the existing capacity is fully utilized, provided the ethylene crackers have more than a 70% operating rate; that is, about 518,000 tons butadiene were extracted in the FRG in 1985.

Butadiene extraction in Cologne was terminated in 1986 after the steam crackers of Esso Chemie were closed down. At the same time the ROW capacity was increased to 150,000 t/a butadiene so that the total capacity remained practically unchanged.

Since there is active world trade with butadiene as a much-sought after raw material for synthetic rubber, the world-wide supply must be estimated. The well-established trade, especially between Western Europe and North America, is due to the fact that much less butadiene is yielded in N. America as a co-product than the demand there. This comes from the differing production structure — the ethylene crackers are fed mainly with ethane because the natural gas fields are richer in that gas (Fig. 2.7). The crack gases contain too little butadiene to make extraction attractive. Processes of dehydrogenation of butane and butenes are energetically more favourable than other syntheses of butadiene e. g. with acetylene as raw material, so that they serve to satisfy temporarily occurring demand peaks in certain markets ("on-purpose-production").

Table 2.11 contains world butadiene capacities and production data for 1983. Of the good 5 million t/a of butadiene produced, about 4.7 million was as co-product from steam cracking and about 0.3 million from dehydrogenation.

Expansion of butadiene production in South America and Eastern Europe would lead first to a bottleneck at the extraction installations; in the other regions the extraction capacities are more than comfortably adapted to the capacities of steam cracker coproduct output and of dehydrogenation.

The average contract prices paid for butadiene in comparison to ethylene and propene during recent years can be seen in Fig. 2.8.

Nowadays ethylene production in the USA is increasing from higher-boiling raw materials. Home supply with butadiene from steam crackers is thus growing and export prospects to the USA of butadiene from other countries are becoming more and more limited. Quantitative forecasts about this are being intensely discussed [2.14].

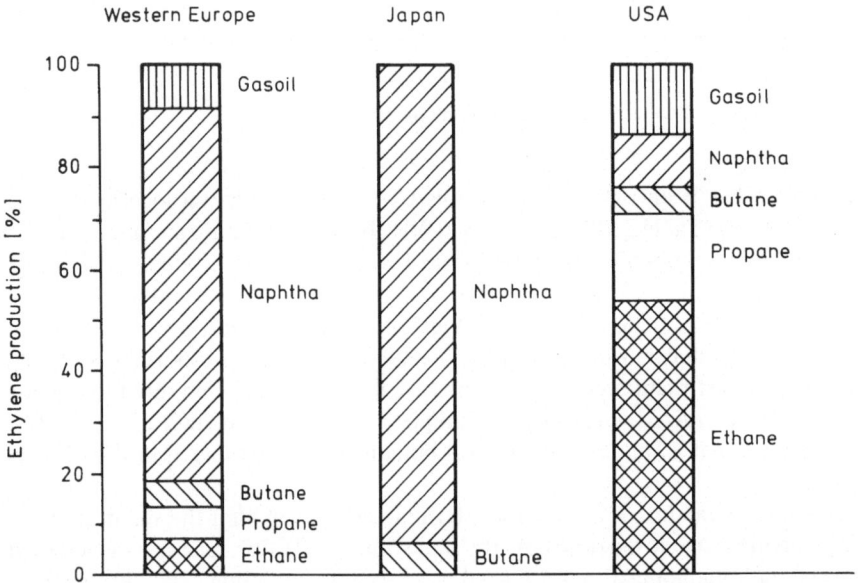

Fig. 2.7. Raw materials for ethylene production in Western Europe, in Japan and in the USA 1983

Table 2.11. World butadiene capacities and production, 1983 in 1000 t/a [2.15]

	Capacities					Production	
	Steam cracker co-prod.	Dehydro-genation	Total	% of Dehydr. Cap.	Extrac-tion	Amount	% of Dehydr.* Prod.
North America	1811	282[1]	2093	13	2450	1200	0
South America	187	95[2]	282	34	252	206	9
Western Europe	2040	0	2040	0	2147	1741	0
Eastern Europe	788	950[3]	1738	55	1115	1065	26
Others	883	45[4]	928	5	1103		0
Totals	5709	1372	7081	Ø 19	7067	5027	Ø 6

[1] USA
[2] Mexico, Argentina
[3] GDR, Poland, Rumania, CSSR, Soviet Union
[4] China, South Africa
* Assumption: Dehydrogenation is operated only when there is an insufficient butadiene supply as steam cracker co-product in the region.

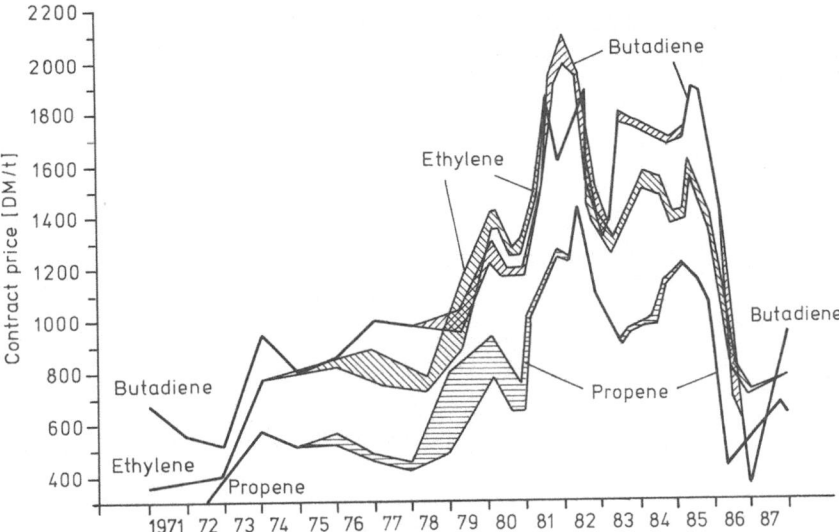

Fig. 2.8. Contract prices for ethylene, propene and butadiene in Western Europe [2.16]

2.2.4 C₄-Hydrocarbons Through Synthesis

Independent syntheses of C_4-base chemicals play only a small part in a situation in which, as at present, petrochemical co-production leads to surplus amounts of C_4-hydrocarbons. However, large quantities of C_4-alcohols are based on propene (see Sect. 2.3). In case this surplus supply of the C_4-hydrocarbons should change, the most important petrochemical syntheses for them from non-C_4-hydrocarbons are given here:

At present, however, the required transformation, as required, of petrochemical base chemicals within the C_4-group is sufficient; this is discussed in more detail in Chapter 3. If there were a change in this situation of surplus, e. g., if petrochemical raw materials were in short supply, the possibilities are mentioned here of how C_4-hydrocarbons may be prepared synthetically using reactions based on coal chemistry or biochemistry.

2.2.4.1 Petrochemical Syntheses

The petrochemical synthetic routes to C_4-base chemicals are shown in Fig. 2.9.

For a few years, particularly in the USA, n-butenes have been yielded as co-products during the production of α-olefines. A broad spectrum of C_4-C_{24}-olefines with an even number of C atoms and a terminal double bond is yielded from ethylene by the Ziegler oligomerisation and modified processes. The C_{12}-C_{18}-fraction is the preferred cut for detergent manufacture; the C_6-C_{10}-fraction is used to prepare plasticisers via OXO-alcohols. The producers of α-olefines in the USA are Ethyl Corp., Pasadena, TX.; Gulf Oil Chemicals, Cedar Bayou, TX.; and Shell, Geismar, LA; in Great Britain, Shell, Stanlow; and in Japan, Mitsubishi. Their total capacity is about 1 million t/a, corresponding at the most to 80,000 t/a 1-butene. All the

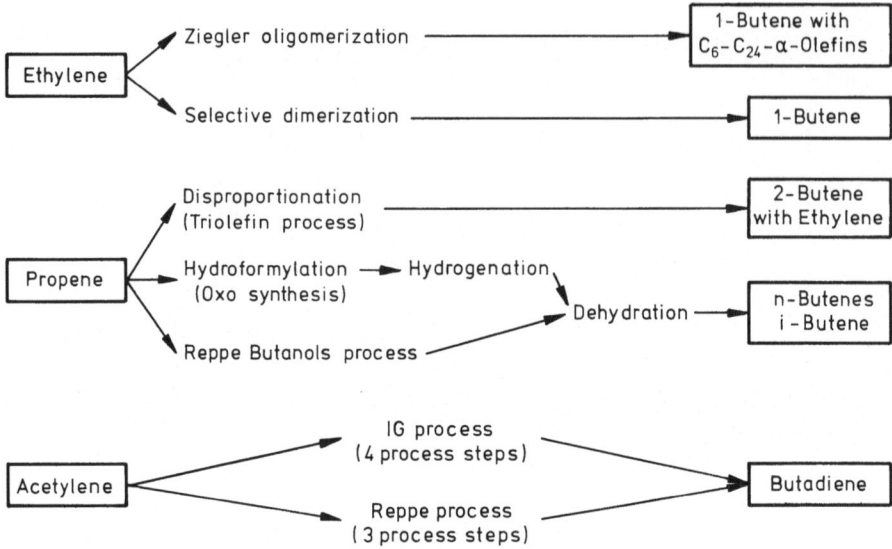

Fig. 2.9. Petrochemical production routes for C$_4$-chemicals

producers are striving, however, to maximise the yield of C$_6$-C$_{20}$-olefines, so that the C$_4$-cut will be processed internally, e. g. Shell with the SHOP-process, by dispropor-tionation to internal olefines [2.17].

A more special petrochemical synthesis is the selective dimerisation of ethylene to very pure 1-butene. This is a further development of Ziegler oligomerisation but using selective complex catalysts containing tantalum or niobium. The reaction proceeds at 80 °C and about 100 bar in the absence of water and of oxygen [2.18]. IFP (Institut Français du Pétrole, Rueil-Malmaison) has developed this synthesis under the name of the Alphabutol process [2.19]. In the meantime, two plants have begun operation under license: in summer 1987 in Thailand and in autumn 1987 a 50,000 t/a 1-butene plant at Arabian Petrochemical (PETROKEMYA) [2.19] in Al Jubail, Saudi Arabia. Furthermore, a license has been awarded to Indian Petrochemicals [2.20] for a 15,000 t/a plant which is to be built in Nagothane, south of Bombay [2.21]. The procedure offers economic advantages on these sites since there are insufficient C$_4$-amounts available there from steamcrackers and the purchase, transport and storage of 1-butene would be highly expensive. Figure 2.10 shows a flow chart of the procedure.

Propene can be converted into ethylene and 2-butene by disproportionation in the Triolefine process of Phillips-Petroleum [2.22].

$$2CH_3-CH=CH_2 \longrightarrow CH_3-CH=CH-CH_3 + CH_2=CH_2$$

The reaction was discovered in 1964 and takes place at 160 °C using a CoO/MoO$_3$-catalyst on a carrier of aluminium oxide. In both the chemical and refinery fields the process can contribute to an economic improvement, provided that propene is available at a very low price. Combination with a dehydrogenation plant, in which the 2-butene is subsequently converted to butadiene, offers the possibility of processing the propene to ethylene and butadiene.

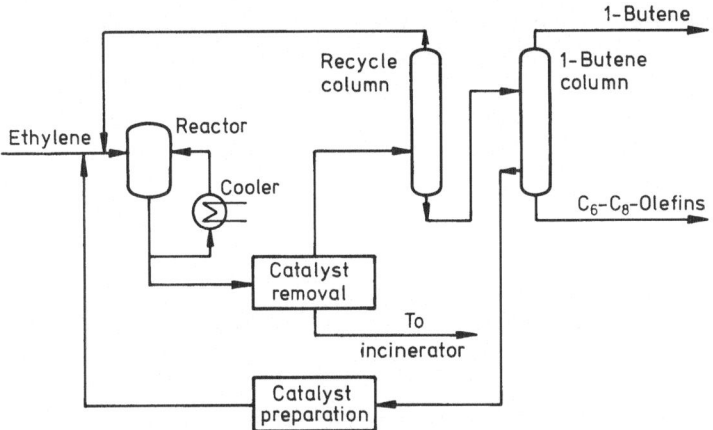

Fig. 2.10. IFP Alphabutol process for the production of 1-butene from ethylene [2.19]

Shawinigan Chemicals, Varennes/PQ in Canada, built a plant for this special use of propene, the only one so far for propene disproportionation. It began production in 1966 [2.23]. However, the market price of propene rose in the 1970s since it was becoming a much sought after chemical raw material and the plant had to be closed down in 1972. As the availability of reasonably-priced propene becomes less, the reverse process is gaining attention, namely, the conversion of ethylene and butenes into propene (see Sect. 3.3.2.3).

The possibility of using the Triolefine process as a prestep before alkylation for gasoline production is also being discussed. Steamcracker propene is then first disproportionated into ethylene and 2-butene and this mixture is reacted with i-butane to "diisopropyl" (DIP) and 2-butene-alkylate [2.22]. At a loss of only 3.7% of alkylate yield, this saves 35% i-butane input in comparison to propene alkylate, with simultaneous improvement in the octane number. Since steam cracker propene could be better processed in the meantime as a chemical raw material, no refinery has yet decided to integrate this process. A preference for the process in the event of increased availability of C_3-fractions from catalytic cracking processes is handicapped by the fact that only pure olefine fractions can be reacted by disproportionation. Refinery propene cuts are thus usually dimerised to high-octane gasoline components (Dimersol G process of IFP) or polymerised (polymer gasoline process).

The butanol syntheses, starting from propene (OXO-synthesis with hydrogenation of the butanals, Reppe butanol process) are of large-scale technical significance; this is discussed more thoroughly in Sect. 2.3.2. The butanol mixture yielded can be easily fractionated into i- and n-butanols. The individual butanols can be dehydrated to the butenes technically but at present this is economically uninteresting. i-Butanol can be dehydrated to i-butene at 300 to 350 °C on an aluminium oxide catalyst. The conversion is practically quantitative, with an i-butene selectivity exceeding 90% [2.24]. n-Butanol can be dehydrated under the same conditions to a mixture of 1- and 2-butenes, although di-n-butyl ether may also be formed at low temperatures. The suggestion has been made of an alternative, namely, selective dehydration of the unfractionated mixture of i- and n-butanols on molecular sieves (calcium zeolite 5Å); only the n-butanol is then dehydrated [2.25].

Acetylene is produced by pyrolysing gaseous or liquid hydrocarbons (BASF acetylene process; Hüls electric arc-plasma process; Hoechst HTP process). At present, the acetylene processes based on these petrochemical sources show financial advantages over those based on coal in the industrial countries [2.26]. However, the synthesis of butadiene from acetylene using the IG- or Reppe-process is, at the moment, uneconomic.

2.2.4.2 Syntheses Based on Coal

Figure 2.11. shows the synthetic routes based on coal.

The synthetic routes based on coal are liquefaction, gasification or acetylene production. Coal gasification as a source of hydrogen and, if necessary, also formaldehyde via methanol, belongs to the acetylene scheme. All the routes have high raw material and energy consumption and, at the present level of coal prices, the routes are much more expensive than the petrochemical alternatives. Should there be a shortage of petrochemical raw materials and sufficient demand for the production of

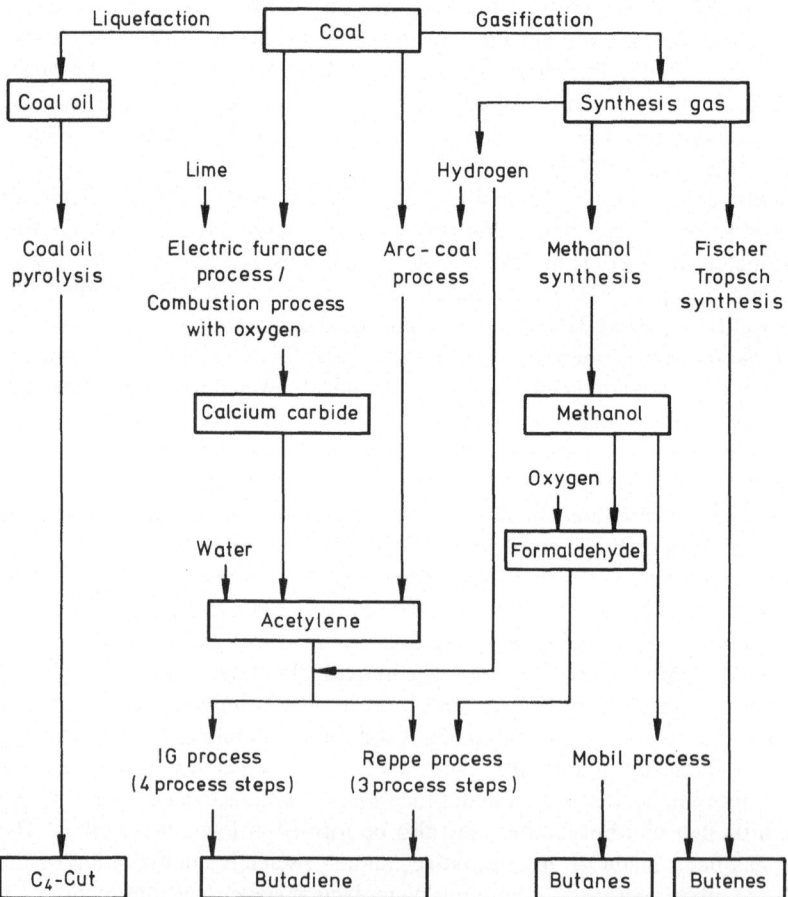

Fig. 2.11. C$_4$-Hydrocarbon synthetic routes from coal

butadiene, the classical acetylene route would probably be comparatively the most favourable one.

Butadiene can be prepared in four stages from acetylene using the Aldol or IG process [2.27].

$2CH\equiv CH \xrightarrow{+2 H_2O} 2 CH_3CHO$ Acetaldehyde

$2 CH_3CHO \longrightarrow CH_3-CHOH-CH_2-CHO$ Acetaldol

$CH_3-CHOH-CH_2-CHO \xrightarrow{+H_2} CH_3-CHOH-CH_2-CH_2OH$ 1,3-Butanediol

$CH_3-CHOH-CH_2-CH_2OH \xrightarrow{-2H_2O} CH_2=CH-CH=CH_2$

As an alternative the three-step Reppe process comes into consideration [2.27].

$CH\equiv CH \xrightarrow{+2HCHO} CH_2OH-C\equiv C-CH_2OH$ 1,4-Butinediol

$CH_2OH-C\equiv C-CH_2OH \Big\langle$
   $\xrightarrow{+H_2} CH_2OH-CH=CH-CH_2OH$ 1,4-Butenediol
   $\xrightarrow{+2H_2} CH_2OH-CH_2-CH_2-CH_2OH$ 1,4-Butanediol

$CH_2OH-CH_2-CH_2-CH_2OH \Big\langle$
   $\xrightarrow{-H_2O} \begin{matrix} CH_2-CH_2 \\ | \qquad \quad \rangle O \\ CH_2-CH_2 \end{matrix}$ Tetrahydrofuran
   $\xrightarrow{-2H_2O} CH_2=CH-CH=CH_2$

In Germany during the very active period of acetylene chemistry, up to 150,000 t/a butadiene was produced using these two processes during the second world war. Although the Reppe process requires only half as much acetylene as the Aldol process, it has also become uneconomical because of its energy requirements compared to the petrochemical routes. At the moment the Reppe process is applied only as far as the 1,4-butanediol stage; in Japan, 1,4-butanediol is being prepared from butadiene in the reverse reaction (see Sect. 3.4.4).

If gasoline production based on coal, via methanol using the Mobil process, succeeds on a large scale there is likely to be considerable production of butane-rich LPG. About 75% by weight of gasoline and 23% by weight of C$_3$/C$_4$-LPG, along with very small amounts of C$_1$/C$_2$ gas are formed when methanol reacts on an acidic zeolite ZSM (zeolite Socony Mobil)-5 catalyst [2.28]. At present, the Mobil process is being used on a natural gas basis and hence more interesting as a petrochemical procedure. Three process variants can be distinguished: MTG (methanol to gasoline); MTO (methanol to olefines); and MTA (methanol to aromatics). The first large-scale plant in this series is the MTG-plant in New Zealand which began production at the end of 1985 [2.29]. About 600,000 t/a gasoline are obtained there from 1.5 million t methanol, including an alkylation using the LPG fraction. A flow chart with balance sheet data based on the methanol charge, is given in Fig. 2.12. The low molecular C$_2$-C$_4$- olefines can be obtained in high yields using the MTO process. The yields of butenes can be increased to over 36% by employing phosphorus-modified ZSM-5 catalysts and a temperature of 400 °C [2.30, 2.31].

Finally, butenes can be derived from coal via the Fischer-Tropsch syntheses with their fixed bed/circulating fluidised bed/slurry phase synthesis variants. Since yields depend on the composition of the synthesis gas, the catalysts and the reaction conditions, representative figures of the yields of C$_4$-hydrocarbons cannot be given. All runs of the experimental plants and of the sole production installation, in

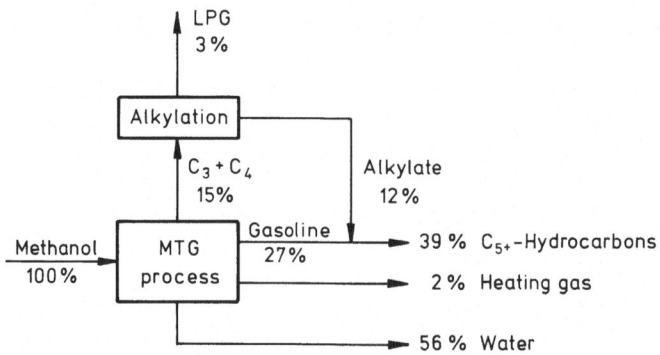

Fig. 2.12. Material balance for the MTG process of Mobil Oil [2.32]

Table 2.12. Butene-selectivity of new Fischer-Tropsch catalysts [2.33]

Catalyst	CO/H$_2$ conversion (%)	Butene-selectivity (%)	Butene yield (%)	of which	
				1-Butene (%)	2-Butene (%)
Precipitated Fe-catalyst (Kölbel)	58	10.4	18	55	45
Fused Fe-catalyst (Sasol)	67	11.2	17	93	7
Fe-catalyst (Ruhrchemie)	34	12.1	36	78	22
Fe-whisker catalyst (Vielstich/Ritschel)	69	12.8	19	93	7
Fe/Mn-catalyst (Kölbel)	50	17.9	36	65	35

Sasolburg, South Africa, indicate that the relatively high C-selectivities are attained for the C$_3$- and C$_4$-olefines but not ethylene. Table 2.12 contains a compilation of butene yields obtained from experimental runs with maximum C$_2$-C$_4$-alkene selectivity. Reference is made here to the literature for information about the individual reaction conditions [2.33].

Little success has been achieved so far in efforts to adjust the selectivity towards ethylene so that the FT-synthesis would fit better into the existing chemistry infrastructure. However, the strength of the process is in the synthesis of straight-chain 1-alkenes of medium and higher C-content. If coal were the source of raw material, the FT-synthesis would be suitable for producing diesel oil. A structural change of chemistry in the direction of C$_4$-hydrocarbons would endow the FT-synthesis with good prospects as a producer of C$_4$ materials.

2.2.4.3 Biotechnological Syntheses

The present viable biochemical routes to C$_4$-chemicals are shown in Fig. 2.13.

Molasses and corn starch can be converted to fuel oil through the butanol-acetone-fermentation process; this is a mixture which contains mainly *n*-butanol and

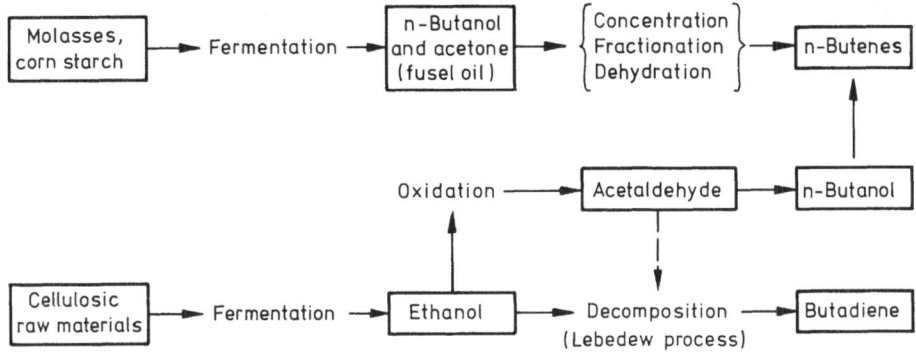

Fig. 2.13. Biotechnological processes for C$_4$-chemicals

acetone [2.34]. After energy-consuming concentration and separation by distillation, *n*-butanol can be dehydrated to *n*-butenes (see Sect. 2.2.4.1).

The Lebedew procedure is the key process for obtaining butadiene via a biochemical route [2.35]. In the first stage ethanol is obtained by fermentation of materials which have undergone saccharification. The ethanol is then decomposed on a hot aluminium oxide catalyst according to:

$$2CH_3CH_2OH \longrightarrow CH_2=CH-CH=CH_2 + 2H_2O + H_2$$

In an improved version of this procedure, a part of the alcohol has to be oxidised to acetaldehyde before the mixture reacts according to

$$CH_3CH_2OH + CH_3CHO \longrightarrow CH_2=CH-CH=CH_2 + 2H_2O$$

to yield butadiene.

After oxidation to acetaldehyde the ethanol can be further processed to *n*-butanol, which can be dehydrated to *n*-butenes.

2.3 Production of C$_4$-Aldehydes and Alcohols

The C$_4$-aldehydes and alcohols are important compound classes alongside the C$_4$-hydrocarbons. A compilation of all the relevant aldehydes, alcohols and diols is given in Table 2.13.

The scope of this study covers materials which can be prepared from C$_4$-hydrocarbons. Other compounds are interesting because butenes can be made from the C$_4$-aldehydes by hydrogenation to the corresponding alcohols and dehydration of the monoalcohols, and butadiene likewise from the diols. For this reason the syntheses of the aldehydes, although really derived from C$_2$- and C$_3$-sources, are briefly discussed.

Table 2.13. Nomenclature and structure of the C$_4$-aldehydes and alcohols

Nomenclature	Abbreviated name	Empirical formula	Structural formula
Aldehydes		C$_4$H$_8$O	
1-Butanal	*n*-Butanal		CH$_3$–CH$_2$–CH$_2$–CHO
2-Butanal	*i*-Butanal		(CH$_3$)$_2$CH–CHO
Alcohols		C$_4$H$_{10}$O	
1-Butanol	*n*-Butanol		CH$_3$–CH$_2$–CH$_2$–CH$_2$OH
2-Methyl-1-propanol	*i*-Butanol		(CH$_3$)$_2$CH–CH$_2$OH
2-Butanol	*sec*-Butanol (SBA)		CH$_3$–CH$_2$–CHOH–CH$_3$
2-Methyl-2-propanol	*tert*-Butanol (TBA)		(CH$_3$)$_3$COH
Diols			
1,4-Butanediol	1,4-Butanediol	C$_4$H$_{10}$O$_2$	CH$_2$OH–CH$_2$–CH$_2$–CH$_2$OH
2-Butene-1,4-diol	1,4-Butenediol	C$_4$H$_8$O$_2$	CH$_2$OH–CH=CH–CH$_2$OH
2-Butyne-1,4-diol	1,4-Butynediol	C$_4$H$_6$O$_2$	CH$_2$OH–C≡C–CH$_2$OH
cis-2,3-Butanediol	} 2,3-Butanediol	C$_4$H$_{10}$O$_2$	CH$_3$–CH(OH)–CH(OH)–CH$_3$
trans-2,3-Butanediol	} (2,3-Buteneglycol)	C$_4$H$_{10}$O$_2$	CH$_3$–CH(OH)–CH(OH)–CH$_3$
1,3-Butanediol	1,3-Butanediol	C$_4$H$_{10}$O$_2$	CH$_2$OH–CH$_2$–CHOH–CH$_3$
1,2-Butanediol	1,2-Butanediol (1,2-Buteneglycol)	C$_4$H$_{10}$O$_2$	CH$_2$OH–CHOH–CH$_2$–CH$_3$
2-Methyl-1,2-propanediol	*i*-Butanediol (*i*-Buteneglycol)	C$_4$H$_{10}$O$_2$	(CH$_3$)$_2$COH–CH$_2$OH

2.3.1 Butanals

The most important process for deriving butanals is hydroformylation (OXO-synthesis) of propene. This was developed by Ruhrchemie in 1938, and involves treating propene with about equimolar amounts of carbon monoxide and hydrogen on metal-carbonyl catalysts at 90 to 180 °C and 25 to 350 bar. This yields a mixture of the butanals; the ratio of *n*-: *i*-isomers can be regulated from 1:1 to 20:1 by choice of catalyst and reaction conditions. [2.36]:

$$\text{CH}_3\text{–CH}_2\text{=CH}_2 \xrightarrow{+\text{CO}\cdot\text{H}_2} \text{CH}_3\text{–CH}_2\text{–CH}_2\text{–CHO} \quad \text{or} \quad (\text{CH}_3)_2\text{CH–CHO}$$

The yield is 90 to 98%. The highest possible proportion of *n*-butanal is aimed to match existing consumption patterns. Recently this has been accomplished with modified Rh-catalysts at low pressure. This reduces the specific consumption of raw material per t *n*-butanal to 0.66 t propene [2.37].

Up to the middle of the 1950s the multiple stage *n*-butanal production from acetaldehyde via crotonaldehyde dominated [2.36]:

$$2CH_3CHO \xrightarrow{-H_2O} CH_3-CH=CH-CHO \xrightarrow{+H_2} CH_3-CH_2-CH_2-CHO$$

Acetaldehyde can be obtained by the addition of water to acetylene, Wacker oxidation of ethylene or dehydrogenation of ethanol. The specific raw material requirement per t *n*-butanal is then 0.854 t acetylene, 0.962 t ethylene or 1 t ethanol [2.37].

In the meantime because of its overall inferior yields, this synthesis has been completely ousted by the OXO-synthesis; it could, however, become important again in countries (e.g. Brazil) where ethanol is obtained from biomass.

Present world production of butanals is of the order of 4 million t/a, of which 3.2 million t/a are *n*-butanal and 0.8 million t/a *i*-butanal [2.36]. About 758,000 t butanals were produced in the FRG in 1986 in the large OXO-plants of BASF, Hüls and Ruhrchemie.

2.3.2 Butanols

Of the four simple butanol isomers, *n*- and *i*-butanol (NBA and IBA) are obtained by hydrogenation of the corresponding aldehydes, or directly using the Reppe butanol process. In a few cases *n*-butanol is also obtained through the butanol-acetone fermentation route which yields fusel oil from molasses or corn starch [2.38].

Up to 1952 *i*-butanol was also obtained by BASF as co-product in remarkable percentages (11 to 14%) from the high pressure methanol synthesis via synthesis gas when using KOH-modified catalysts [2.39]. Numerous procedures for preparing mixtures of methanol and higher alcohols, including butanols, are presently under development (so-called fuel methanol). The reader is referred to the special literature for further information about these procedures of IFP, Lurgi, Dow Chemical, Chem Systems and the MAS (methanolo + alcooli superiori) process of Snamprogetti-Topsøe-Anic [2.40]. As an example showing the possible composition and the properties of the product, the data for the MAS process are given in Table 2.14; the large fraction of C$_4$-alcohols is noteworthy.

Table 2.14. Composition and properties of MAS [2.41]

Component	weight %
C$_1$alcohol	68–72
C$_2$alcohol	2–3
C$_3$alcohol	3–5
C$_4$alcohol	10–15
C$_5$alcohol	7–12
ketones and aldehydes	max. 2
water	max. 0.1

density at 20 °C 0.804
octane blending values RON 120–135; MON 93–106;
(RON + MON)/2 106–121

n- or *i*-Butanals are generally hydrogenated in the gas phase on Ni- or Cu-carrier catalysts at temperatures of 115 to 160 °C and pressures of 2.5 bar. Conversion is almost quantitative so that there are few by-products [2.42].

The Reppe butanol process differs from the OXO-synthesis route in that propene is converted with 90% yield directly at low temperature (100 °C) and low pressure (5 to 20 bar) to the butanols in the *n* : *i* ratio of 86 : 14 [2.43].

$$CH_3 - CH=CH_2 \xrightarrow{+3CO+2H_2O} \begin{cases} CH_3-CH_2-CH_2-CH_2OH + 2CO_2 \\ CH_3 \\ CH_3{\Large\diagup}CH-CH_2OH + 2CO_2 \end{cases}$$

Despite its advantages (no hydrogen needed, favourable *n* : *i* ratio) the procedure has found only a single operator, Japan Butanol (30,000 t/a). The reasons are to be found in the high consumption of catalyst and solvents and, in comparison with the OXO-process, more elaborate production techniques and lower investment cost degression related to capacity increase. About 313,000 t/a *n*-butanol and 80,000 t/a *i*-butanol were produced in the FRG in 1986, exclusively via OXO-synthesis/hydrogenation [2.44].

The hydration of *i*-butene yields *tert*-butanol (TBA). The addition of water can be performed indirectly in the presence of sulfuric acid, or, in more modern processes, directly on ion exchangers. *sec*-Butanol (SBA) can also be produced indirectly from *n*-butenes but only at a higher concentration of acid than for hydrating *i*-butene. A direct procedure for SBA according to the ion exchanger principle has recently been developed by Texaco and incorporated into a commercial plant.

The raw material basis for TBA is usually raffinate I, for SBA, raffinate II. Since these processes form the first steps towards raffinate separation they are treated in further detail in Sects. 3.1.2.2 and 3.1.2.4.

The market leader for TBA is the Atlantic Richfield Corp. (Arco), with a capacity of approx 1.36 million t/a. The production sites, erected by Oxirane, a former joint venture of Arco and Halcon, are concentrated in Bayport, TX (880,000 t/a TBA) and Botlek, near Rotterdam (480,000 t/a TBA). From 1988, the Arco plant in Fos-sur-Mer near Marseilles, at present under construction, with 400,000 t/a TBA, will be added. In all the plants, TBA is yielded exclusively as a co-product of the Oxirane process [2.45]:

$$\begin{array}{ccccc}
CH_3 & & CH_3 & & CH_3 \\
CH_3{\Large\diagdown}CH & \xrightarrow{+O_2} & CH_3{\Large\diagdown}C-O-O-H & \xrightarrow{+CH_3-CH\cdot CH_2} & CH_3{\Large\diagdown}C-OH \quad + \quad CH_3-CH-CH_2 \\
CH_3{\Large\diagup} & & CH_3{\Large\diagup} & & CH_3{\Large\diagup} \qquad\qquad\quad \diagdown\!O\!\diagup
\end{array}$$

$$\begin{array}{cc}
\xrightarrow{+H_2} & CH_3 \\
 & CH_3{\Large\diagdown}C=CH_2 \xleftarrow{-H_2O}
\end{array}$$

i-Butane is converted to a peroxide in the liquid phase at 120 to 140 °C and 24 to 35 bar. The peroxide decomposes spontaneously into *tert*-butanol and small amounts of acetone and carboxylic acids. At 25 to 50% *i*-butane conversion, hydroperoxide yields of 50 to 60% are obtained. The side products formed are removed with the exhaust from the peroxide formation. About 0.4 m³ of waste water is formed per t propene oxide (PO). It contains components which are easily degraded biologically. Non-reacting *i*-butane is distilled off. After adding a molybdenum catalyst the hydroperoxide reacts at maximal 180 °C with propene which is introduced in excess. The hydroperoxide is converted completely. PO is formed with about 98% selectivity with reference to reacted propene. The liquid reaction product is further processed by multiple stage distillation; unconverted propene is recycled to the plant.

The amount of propene oxide can be varied by recirculating part of the TBA in order to dehydrate it catalytically to *i*-butene and re-hydrogenate it to *i*-butane. Without carrying out circulation, which is an advantage economically, about 2.5 t TBA are yielded per ton of propene oxide. The purity of the TBA reaches 95%. Impurities of this so-called "gasoline grade" TBA have been mentioned [2.46]: 2.5% by weight butanes and 1% by weight acetone, 0.5% others (SBA, MEK, carboxylic acids). The plant at Botlek obtains the raw material *i*-butane mainly from the LPG-fractionation plants in Teesside, England and from Karst, Norway. *n*-Butane is also delivered in special tankers from refineries on the coast in the neighbourhood; it is isomerised in the butamer installation on the site.

TBA can also occur as a co-product during all MTBE processes when water is present in the reaction mixture. The production of MTBE is treated in Sect. 3.3.1.2.

Capacities of 60,000 t/a SBA and 50,000 t/a TBA as chemical intermediates existed in the FRG in 1985.

2.3.3 Butanediols

As far as quantity is concerned, 1,4-butanediol is the most important C$_4$-diol. At present it is prepared almost entirely from acetylene using the Reppe synthesis (see Sect. 2.2.4.2). Butynediol and butenediol are thereby yielded as intermediate, unsaturated compounds. Small amounts of these are utilized for the preparation of crop protection agents and other specialities; for example, Solvay prepares a brominated derivative of butynediol which is added to plastics as a flame retardent.

The acetylene route has a high energy consumption. Consequently new syntheses are being investigated, especially in Japan. Methods of preparing 1,4-butanediol from butadiene are presented in Sect. 3.4.4. In the future the route via maleic anhydride (MA) could be of interest; MA can be derived from benzene, *n*-butane (see Sect. 3.2.3) or raffinate II (see Sect. 3.3.3.4). Maleic anhydride can be directly, or after hydration to maleic acid, hydrogenated on suspended NiCrMo-catalysts to a mixture of tetrahydrofuran (THF), γ-butyrolactone and 1,4-butanediol [2.47]. Two Japanese companies are applying this process (see Sect. 3.4.4). The yield is centred on THF in fact, but this compound can be converted to 1,4-butanediol by hydrolysis in the presence of sulfuric acid.

Standard Oil intends to produce 1,4-butanediol, γ-butyrolactone and tetrahydrofuran from *n*-butane in the USA in 1989 [2.48]. The *n*-butane is to be oxidised to maleic anhydride (about 100,000 t/a). MA is then esterified to diethyl maleate which is also an intermediate for the insecticide Malathion. The ester will be converted to the above-mentioned products on copper catalysts at 120 to 140 °C and 14 to 42 bar. Since the relative amounts of these products can be varied, there might be some economic advantages over the acetylene route.

A new 1,4-butanediol process starts from propene. It belongs to Daicel, who have had a 60,000 t/a plant in Japan since 1983 [2.49].

$$CH_2{=}CH{-}CH_3 \xrightarrow[-HCl]{+Cl_2} CH_2{=}CH{-}CH_2Cl \xrightarrow[-HCl]{+H_2O} CH_2{=}CH{-}CH_2OH \xrightarrow{+CO_2 \ +2H_2}$$

$$CH_2OH{-}CH_2{-}CH_2{-}CH_2OH$$

          Allyl chloride   Allyl alcohol

Nowadays 2,3-butanediol is made from raffinate II, via 2,3-butene oxide (see Sect. 3.3.3.6). In the USA it was also made formerly by fermentation of pentoses and hexoses.

1,3- and 1,2-Butanediols have remained relatively unimportant economically up to now.

i-Butanediol can be made from pure i-butene directly by a catalytic redox procedure or indirectly via i-butene oxide. It can be converted to MMA via α-hydroxyisobutyric acid (see Sect. 3.3.4.3).

Capacities are available in the FRG for 1,4-butanediol made by the Reppe-process, of 110,000 t/a at BASF and 70,000 t/a at GAF Hüls Corporation, Marl.

3 Processes of Separation and Transformation in C₄-Chemistry

3.1 Separation of the C₄-Cut and C₄-Gases from Refineries

Pure butadiene cannot be separated from the C_4-cut by distillation. Both 1,3-butadiene and *trans*-2-butene form azeotropes with *n*-butane; vinylacetylene likewise forms azeotropes with *cis*- or *trans*-2-butenes, differing only slightly in boiling point from butadiene. Finally, the boiling points of *i*-butene and 1-butene are so near that their distillative separation is uneconomic (Table 3.1).

Butadiene is thus separated through selective chemical complex-formation with an auxiliary, followed by its decomposition; or mostly by using selective solvents and then separating the extract by distillation.

The butadiene-free C_4-cut (raffinate I) is separated by exploiting the different reactivities of *n*- and *i*-butenes towards sulfuric acid; or particular butenes or butanes are removed selectively with molecular sieves and the residue is then processed by distillation. Other procedures exist, based on the differing reactivities of the components; these include some modern sulfur-free processes.

Separation may also include removal of impurities, e.g. explosive acetylenes or butadiene residues which would make the handling difficult or impair the purity of the product. Careless handling of C_4 acetylenes* led, in 1969 at Union Carbide in Texas City, USA, to an explosion with serious consequences; a butadiene extraction unit with dimethylacetamide as the solvent was completely destroyed [3.2].

Table 3.1. Boiling points of the principal components of the C_4-cut [3.1]

Hydrocarbon	Boiling point in °C	Hydrocarbon	Boiling point in °C
Propene	−47.70	*n*-Butane	−0.5
Propane	−42.07	*trans*-2-Butene	+0.88
Propadiene	−34.5	*cis*-2-Butene	+3.72
Propyne	−23.22	Vinylacetylene	+5.1
i-Butane	−11.73	Ethylacetylene	+8.1
i-Butene	−6.9	Diacetylene	+10.3
1-Butene	−6.26	1,2-Butadiene	+10.85
1,3-Butadiene	−4.41	Dimethylacetylene	+26.9

---

* The concentration of vinylacetylene in the extract must not exceed 35% by volume (explosive limits of the mixture). It is monitored continuously and its magnitude determines the procedural method of the plant. The measuring instrument becomes gradually coated with polymer. It had been dismantled and cleaned on the day of the misfortune but the plant had not been shut down during the maintenance work.

Hydrogenation processes were developed early to remove acetylene fractions from steam cracking fractions. These led to remarkable losses of butadiene as a result of the inadequate selectivity. These processes have been continuously improved through development work on selective cold hydrogenation at Bayer [3.3, 3.4] and also through work at IFP [3.5], Hüls [3.5] and UOP (Unibon process) [3.6]. The Bayer Cold Hydrogenation process is carried out all over the world with a throughput of more than 9 million t/a at some 70 refineries and chemical plants.

In the modern Hüls SHP process, polyunsaturated hydrocarbons in olefine mixtures are added quantitatively to the corresponding monoolefines, without these being hydrogenated to the alkanes or isomerised. This takes place with the hydrogen dissolved in the liquid, on Pd-catalysts in the presence of a few ppm carbon monoxide. The SHP process is the preferred preliminary stage of purification of the major part of the 1-butene worldwide. SHP-installations have been built at Hüls in Marl; Exxon Chemicals in Baytown, TX; Paramins (subsidiary of Exxon) in Baytown, TX; Mitsui in Chiba; and Tonen in Kawasaki.

3.1.1 Separation Processes for Obtaining Butadiene

Physical and chemical processes are distinguished by the procedures for separating butadiene. The former are especially suitable for C$_4$-cuts rich in butadiene and strongly contaminated with acetylenes (propyne, butyne, vinylacetylene, diacetylene) and allenes (propadiene, 1,2-butadiene). Butadiene purity is subject to strong

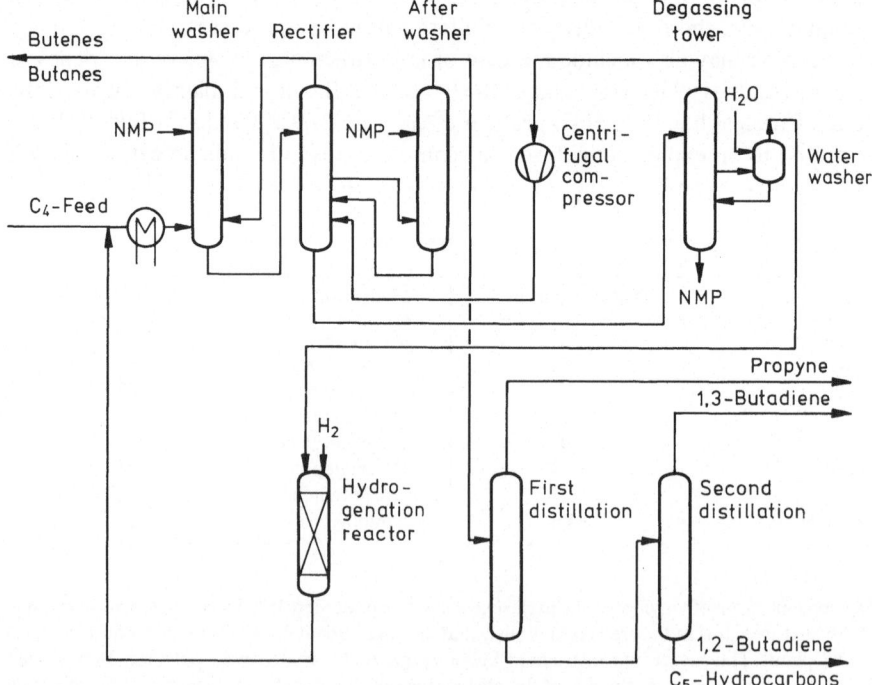

Fig. 3.1. Extractive distillation to obtain 1,3-butadiene with NMP [3.7]

requirements (see Sect. 1.4); hence these impurities must remain completely in the raffinate, or be separated and burnt, or treated by selective hydrogenation (see Sect. 3.1). The physical processes are based on the principle of extractive distillation, using selective solvents. A general process scheme is given in Fig. 3.1 depicting the procedure which varies essentially only in the choice of solvent and technical details.

In order to increase selectivity the extraction agent is introduced in aqueous solution at the head of the main and after washers. All components which are less soluble in the extraction agent than butadiene escape via the head of the main washer and can then be separated into C_3-hydrocarbons and C_4-raffinate I in a simple distillation stage. The after washer ensures that no C_4-acetylenes remain in the product stream, from which, after two distillation stages, pure 1,3-butadiene is obtained at the head of the second distillation column. The dissolved components are withdrawn continuously from the bottom of the rectifier. These are removed in a degasing column, the extraction agent being recovered. At the same time, the C_4-acetylenes are removed in the suitable zone. Mild hydrogenation of the C_4-acetylenes in the hydrogenation reactor is worth while in every case since it yields a butadiene-containing C_4-cut which can be recycled. The C_5-hydrocarbons, 1,2-butadiene and propyne can be hydrogenated to give gasoline components but it is better to use them as fuel to avoid additional risks and costs.

The best known physical procedures for separating butadiene, with the relevant solvents in aqueous solution are:

– the BASF process with N-methylpyrrolidone [3.8]
– the Shell process with acetonitrile [3.9]
– the Phillips process with furfurol [3.10]
– the Nippon Zeon process (formerly Japanese Geon) with dimethylformamide [3.11]
– the UCC (Union Carbide Corp.) process with dimethylacetamide [3.12]

An aqueous solution of cuprous ammonium acetate is used as the complexing agent in the Esso process and has technical advantages for obtaining butadiene from dehydrogenated cuts of only low butadiene content [3.13]. Furthermore, it is an advantage that dehydrogenation cuts contain only very small amounts of acetylenes. There is a tendency for foam to form and interfere in the process if the acetylene content exceeds 100 ppm. Consequently a preliminary treatment of the C_4-fraction by selective hydrogenation is necessary occasionally (see Sect. 3.1). The technical procedure is, however, more complicated than the physical processes on the whole. Thus a minimum of eight mixer/settler units are needed, and recovery of the acetate solution without affecting the environment is technically complicated.

The capacities of the individual procedures in different parts of the world are given in Table 3.2. Section 6.2.2.3 contains information about the costs of butadiene extraction.

Fig. 3.2 shows a butadiene extraction plant using NMP.

Table 3.2. Capacities for butadiene extraction by region and process, in 1000 t/a

Process Extraction agent	BASF NMP	Shell ACN	Phillips FUR	Nippon Zeon DMF	UCC DMA	Esso CAA	Total
North America	0	1016	362	674	127	271	2450
South America	50	95	0	52	0	55	252
Western Europe	870	887	100	232	0	58	2147
Eastern Europe	200	0	255	615	0	45	1115
Remaining world	140	361	0	557	0	45	1103
Total	1260	2359	717	2130	127	474	7067
% share of process	18	33	10	30	2	7	100

NMP = N-methylpyrrolidone; ACN = acetonitrile; FUR = furfurol; DMF = dimethylformamide; DMA = dimethylacetamide; CAA = cuprous ammonium acetate

Fig. 3.2. A butadiene extraction plant with NMP at Bunawerke Hüls GmbH, Marl.

3.1.2 Separation Processes for Obtaining Butenes

3.1.2.1 Survey

In most of the conversion processes of C$_4$-chemistry, the specifications of the feedstocks are subject to special demands. For butanes, the feedstock spectrum is restricted to i-butane, n-butane or a mixture of the two. For butene processes, however, the purity requirements are more markedly different. Either raffinates, butene-concentrates or pure butenes are needed. The preparation of pure butenes is

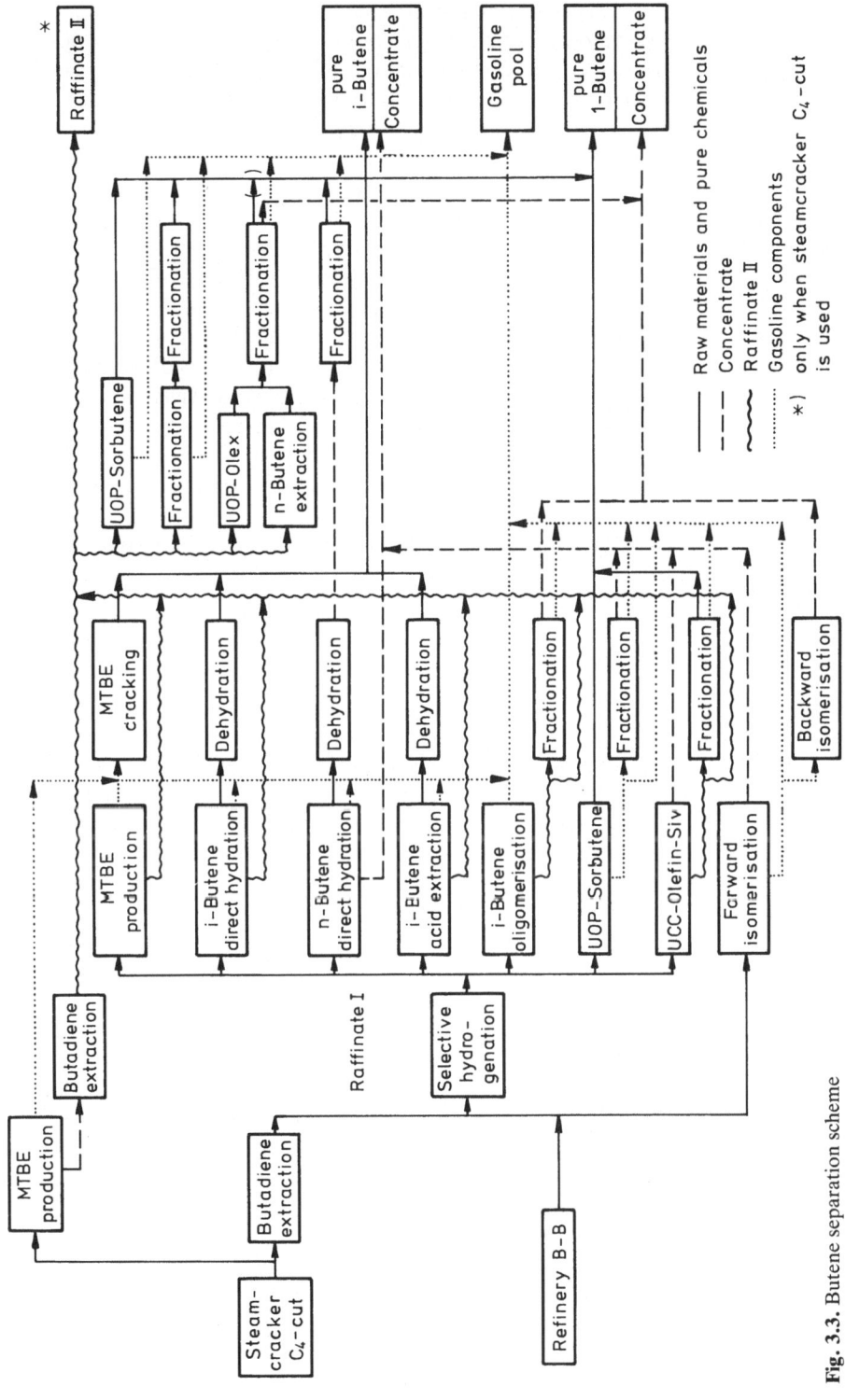

Fig. 3.3. Butene separation scheme

characteristic and in many cases a compulsory condition for C_4-chemistry; therefore a large number of separation processes has been developed. The choice of the best alternative from among this complex array of separation procedures is determined by the need for particular butene isomers; this is fixed by the existing production structure of a chemical factory and may be coordinated with future activities within strategic planning. In order to facilitate the choice of planning alternatives all the relevant routes for separating butenes are clearly arranged in Fig. 3.3.

The separation is carried out by chemical or physical procedures. The chemical procedures are based on selective reactions of the butenes, such as formation of alcohols or ethers, or oligomerisation and isomerisation. The physical procedures depend on the differing adsorption properties of molecular sieves. For fractionation one makes use of boiling point differences or, in the case of extraction, from the solubilities of the components in selective solvents. Most processes of separation are preceded by selective hydrogenation to remove the acetylenes and allenes, unless this has already been performed before the butadiene extraction. If pure n-butene isomers have to be obtained, this separation stage must also be preceded by selective hydrogenation in order to take out butadiene residues (for hydrogenation procedures, see Sect. 3.1).

Information concerning the cost of butene separation is to be found in Sects. 6.2.2.1 and 6.2.2.2.

3.1.2.2 Sulfuric Acid Processes

Extensive development work has led to processes which take advantage of the differing reactivities of the butenes towards sulfuric acid. i-Butene is hydrated to *tert*-butanol (TBA) in the presence of 50 to 60% sulfuric acid even at 0 °C, whereas the reaction of n-butenes to *sec* – butanol (SBA) begins only in 75 to 80% sulfuric acid at 20 to 35 °C. The C_4-cut (raffinate II), freed from i-butene with the sulfuric acid of lower concentration, can be separated by distillation. If desired, the n-butenes can also be removed at the second stage by using the more concentrated sulfuric acid.

Oligomers of i-butene, which cannot be dissociated into the original monomer, are formed at higher acid concentration and low temperature. This oligomerisation is considered to be a troublesome side reaction during preparation of TBA but it can be deliberately carried out as the so-called Cold Acid process if one is dependent on a highly pure raffinate II or wishes to obtain DIB/TIB (see Sect. 3.3.2.2). If the temperature is raised to 100 °C all the butenes react to give a mixture of oligomers which may perhaps go into the gasoline pool. TBA is dehydrated to the chemical raw material, i-butene, with over 90% selectivity, on aluminium oxide in an oven at 250 °C. Provided it finds no use as a solvent or chemical intermediate, surplus TBA can be added to the gasoline pool. Most of the SBA is employed to make methyl ethyl ketone (see Sect. 3.3.3.3) but here, too, suitable dehydration to a mixture of n-butenes is possible.

The following are the most important procedures developed and used on a large technical scale for separating i-butene with sulfuric acid:

– the Esso process, forerunner among procedures for obtaining i-butene. It employs 60 to 65% sulfuric acid, which has to be diluted to 45% before the thermal dehydration [3.14]. Among others a plant was operating at Esso Chemie in Cologne until the naphtha cracker was closed down in 1986.

– the CFR (Compagnie Française de Raffinage) procedure, where the extraction and regeneration are carried out with 50% sulfuric acid [3.15]. This process is utilized at a plant at Polysar, Antwerp.

– the BASF process, using 45% sulfuric acid. The TBA is separated in a vacuum before dehydration, thus furnishing especially pure i-butene [3.16]. ROW is operating a plant in Wesseling.

In many cases hydration with sulfuric acid creates environmental problems [3.17]. The product streams from the process must be freed from entrained drops of sulfuric acid. Waste water containing organic materials and salts is yielded, especially in the first two processes mentioned above. In the BASF process, purification is accomplished by rectification so that there is no effluent except the reaction water from the dehydration. The sulfuric acid is led back into the absorption stage after concentration (in the Esso procedure) or directly (in the other two procedures). Part of the acid has to be removed, however, to limit the build up of acid tar [2.18].

3.1.2.3 Molecular Sieve Processes

Modern molecular sieve extraction can replace sulfuric acid extraction before the C_4-gases are subjected to fractional distillation.

1-Butene, which interferes in the distillation, can be selectively absorbed in the Sorbutene process of UOP (Universal Oil Products, Des Plaines, IL, USA) [3.19]. The 1-butene absorbed is washed out with a desorbing liquid of distinctly different boiling point and then separated from it by distillation. The yield of 1-butene is 92% and the product purity 99.2%.

The OlefinSiv process of UCC, in which all the n-butenes and n-butane are extracted from the C_4-mixture, competes with the above [3.20]. A fractionating column can be placed after the absorber in order to separate 1-butene and i-butane. i-Butene of 99% purity can be distilled from this. 1-Butene in 99% purity can be obtained from the n-C_4-extract using another fractionating unit; the 2-butene and n-butane remain in the bottom and generally do not need to be separated. A C_{8+} fraction which is yielded can be used as co-product or reintroduced into the process. Some isomerisation of 1-butene to 2-butene cannot be avoided. The UCC technology was developed by Linde, a company which, with the Iso-Siv process, already possessed molecular sieve know-how in the field of n/i-paraffin separation.

The Olex process of UOP is a third version of molecular sieve extraction. It enables all the butanes to be separated, so that only olefines remain in the residual stream [3.19]. The most favourable way is to arrange the process for a highly effective separation of i-butene, e.g. through MTBE production, direct hydration, or acid extraction of the i-butene. Under these conditions, pure 1-butene can be obtained after separation of the butenes by distillation; otherwise only a concentrate with the remainder of the i-butene is obtained. This process is particularly interesting for refineries which, for example, by subsequent butane isomerisation, prepare octane improving agents from i-butane. The Olex process could also form the basis of possible future isoprene production through butene disproportionation (see Sect. 3.3.4.4).

3.1.2.4 Other Processes

Recently a direct hydration of the *i*-butene from a steam cracker C_4-cut to TBA using the ion exchanger principle has been achieved on a large technical scale [3.21]. After hydration with the help of an acid ion exchanger TBA dissolves in the C_4-hydrocarbon phase and can easily be separated from the circulated suspension of the ion exchanger. After replacement of the water bound by the hydration, this suspension is recycled. TBA itself functions as a solvent for the *i*-butene/water mixture in the procedure of Hüls, whereas additional acetic acid is added in the Mitsubishi Rayon procedure. The residual C_4-hydrocarbons are separated by distillation, yielding crude TBA containing about 9% water. A flowchart of the process is shown in Fig. 3.4.

Pure TBA (> 99.9% by wt.) can be distilled, or a TBA/H_2O azeotrope obtained, from the crude TBA. The crude material can also be decomposed to extremely pure *i*-butene (99.98% by wt.) at temperatures below 150 °C and slight pressure, with heterogeneous catalysts. All these mentioned process stages have been effected since 1981 in a plant near Hüls; it is able to produce 50,000 t/a pure TBA and up to 30,000 t/a *i*-butene [3.23].

The Adib process for obtaining pure *i*-butene from catcracker C_4-cuts has been tried out in an Argentine refinery [3.24]. In this, *i*-butene from refinery B-B is selectively reacted with phenol in the gas phase using an acid catalyst, to yield all the mono-, di-, and tri- *tert*-butylphenols; these products are easily separated as liquids together with the catalyst. Heating liberates the *i*-butene as a gas of > 99.5% purity, and the phenol can be recycled with the catalyst. Figure 3.5 shows a flowsheet of the continuously operated demonstration plant in Argentina.

A procedure for separating *n*-butenes via *sec*-butanol, without using sulfuric acid, is a recent development of Deutsche Texaco. Direct hydration using the ion exchanger procedure was accomplished for the first time in a 60,000 t/a plant in Moers-Meerbeck [3.25] in 1983, leading to the closing down of the indirect hydration via sulfuric acid (40,000 t/a). It was stressed that there is a better yield of product and lower energy consumption in comparison with the process using the acid. In addition, emissions could then be reduced considerably. Idemitsu Petrochemical in Japan developed a similar procedure which started production (40,000 t/a) in 1985. Figure 3.6 shows a simplified chart of the process which uses a homogeneous dilute catalyst

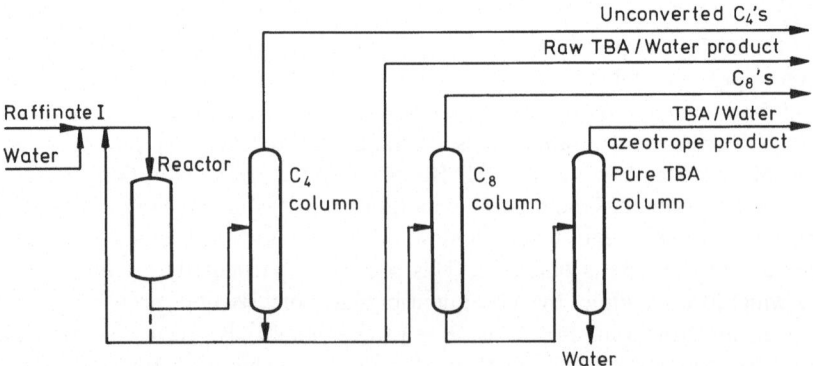

Fig. 3.4. TBA process of Hüls [3.22]

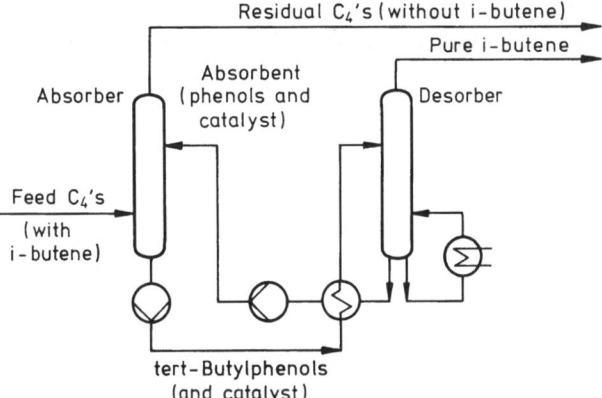

Fig. 3.5. Adib process for extracting *i*-butene [3.24]

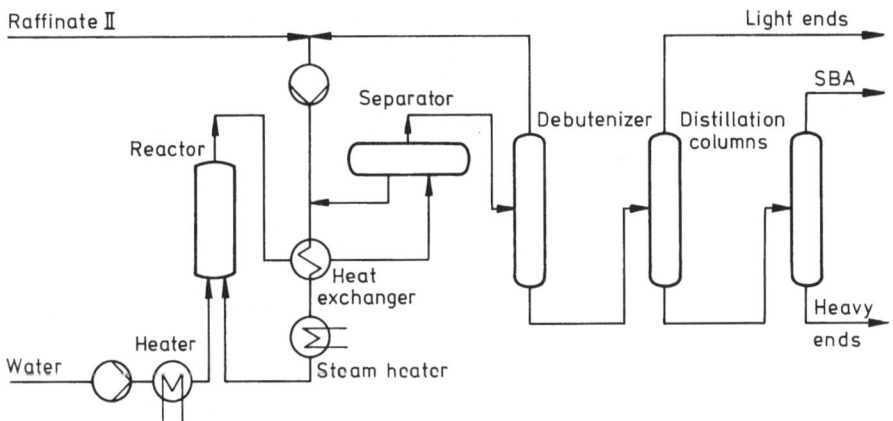

Fig. 3.6. Idemitsu SBA process [3.26]

solution. Within the reactor, supercritical gas extraction is applied. The reaction pressure is 150 to 250 bar, the reaction temperature is between 170 and 230 °C [3.26].

The production of MTBE may also be regarded as a sulfur-free process for separating *i*-butene. The MTBE processes are treated in more detail in Sect. 3.3.1.2. A thermal-catalytic reverse cracking of MTBE has been proposed for sulfur-free production of pure *i*-butene [3.26, 3.27]. Acid-reactive metal oxides or phosphoric acid on a kieselguhr carrier come into consideration as catalysts for this reaction. Methanol is yielded as by-product. Exxon had already reported their intention of following this route after putting into operation the Neochem MTBE process (based on raffinate B-B) at the factory in Baton Rouge, LA in the USA; nevertheless, only the MTBE plant has been constructed, probably because it has become apparent that the required MTBE-purities (99% wt. %) would have impaired the profitability of the plant too much. Sumitomo Chemical in Chiba are running a Hüls MTBE plant based on steam cracker raffinate I and are cracking the pure MTBE to *i*-butene and methanol. 1-Butene is being obtained there from raffinate II, using selective hydrogenation. The capacity for *i*-butene and 1-butene is said to be about 50,000 t/a in each case. In Hungary also there is to be, in the near future, a Snamprogetti MTBE plant with subsequent cracking to furnish 500 t/a *i*-butene for speciality secondary products [3.28].

Another sulfur-free procedure for separating *i*-butene is the oligomerisation process of Bayer (see Sect. 3.3.2.2). The C$_4$-fraction containing *i*-butene is converted to di- and triisobutene (DIB/TIB) in the presence of a cationic ion exchanger. A disadvantage is that there is some conversion of 1-butene into 2-butene. The residual C$_4$-raffinate contains only about 1% *i*-butene. It is, however, not possible to convert the oligomers back to *i*-butene.

A final possibility is to separate *i*-butene and *n*-butenes via isomerisation. After isomerising 1-butene to 2-butene under mild hydrogen pressure (so-called hydroisomerisation), 2-butene and *i*-butene can be separated from each other by extractive distillation, usually with the extraction agent furfurol. Phillips and IFP have based proprietory processes on this method [3.5]. Isomerisation by IFP has been carried out since 1978 at Petro-Tex in Houston, TX. After distillation the purities attained are only 92 to 95% for *i*-butene and 2-butene. The material loss during isomerisation is below 2% for *n*-butenes and less than 0.5% for *i*-butene. These procedures have been able to become established for processing refinery B-B in some refineries because the 2-butene is welcome directly for the gasoline pool, or as a component for alkylate production on account of the relatively high octane numbers of the product.

A simple distillation can be used instead of an extractive distillation after the isomerisation of 1-butene to 2-butene. It enables the content of *i*-butene in refinery B-B to be increased to 27 wt. %, which makes preparation of TBA economic. The General Sekiyu Refineries have adopted this principle; they carry out further processing, based on refinery B-B and processes of Hüls converting *i*-butene to TBA (50,000 t/a) and 2-butene to α-octene (45,000 t/a) [3.22].

A modification of this technique is the two-stage forwards/backwards isomerisation. The forward step corresponds to the isomerisation of 1-butene to 2-butene, already described above. The extractive distillation is followed by the complete reverse isomerisation of 2-butene to 1-butene. There appears to be no technical realisation.

cis-2-Butene can be separated by simple distillation from steamcracker-C$_4$-raffinates and catcracker C$_4$-streams. Shell in Pernis in this way obtains the raw material for producing SBA (85,000 t/a).

The separation of 1-butene is sometimes desired. This can be performed by using all fractions by means of the Sorbutene molecular sieve process of UOP, which yields purities of 1-butene of up to 99.2%. The elimination of high purity 1-butene (Specification is given in Sect. 1.4) is more costly: a high quality raffinate II fraction and at least two separation stages are required. Some isomerisation of 1-butene into 2-butene can occur and cause material losses when using molecular sieve processes. This is especially true for the Olex process which comes into consideration for the present case. It may thus be better to carry out a prior, extractive separation of the *n*-butenes from the butanes in order to avoid this material loss. The same selective solvents can be employed as in the case of butadiene extraction (see Sect. 3.1.1). 1-Butene can be distilled as a pure product from the mixture of *n*-butenes thus obtained and pure 2-

butene is yielded as a co-product. No secondary processes using pure 2-butenes are known so that it is better to separate first i-butane and then 1-butene in two successive fractionations. This procedure is encountered at Hüls [3.30].

Krupp-Koppers is running a pilot plant in Essen for extractive distillation of C_4-components from various C_4-streams. By means of the extraction agent Butenex it was possible to separate raffinate I into the fractions butanes/butenes, and raffinate II into butanes/1-butene/2-butene. $trans$-2-Butene can be largely separated from n-butane but it is not possible to separate i-butene and 1-butene. This extraction agent is said to be a morpholine derivative [3.31]. The pilot plant in Essen is shown in Fig. 3.7.

Fig. 3.7. Pilot plant for extractive distillation of C_4-components with Butenex (Krupp-Koppers)

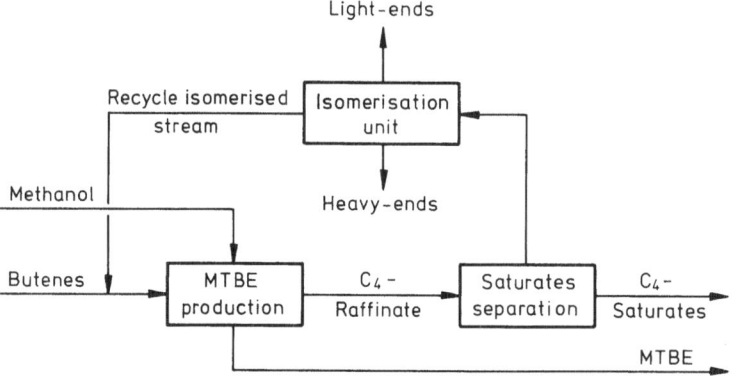

Fig. 3.8. MTBE production by integrated skeletal isomerisation of n-butenes to i-butene [3.31]

The skeletal isomerisation of *n*-butenes to *i*-butene has reached an advanced stage of development. The conversion per pass is 35% and the selectivity with reference to *i*-butene exceeds 80% in the process of Snamprogetti, using an organosilicon/aluminum carrier catalyst [3.31]. Light ends are yielded as co-products, with the quality of C$_3$/C$_4$-alkylate gasoline. Experts are optimistic that the technology will soon reach the commercial stage [3.29]. Figure 3.8 shows the integration of the skeletal isomerisation into a production scheme for MTBE, and when in general operation the MTBE production from steam cracker C$_4$-fractions could be increased by about 80%.

3.2 Processes of Butane Conversion

3.2.1 Isomerisation

Since there are only limited technical possibilities of utilizing *n*-butane there is practical interest in its isomerisation into *i*-butane. The Butamer process of UOP has assumed very great importance in this connection [3.32]. Highly pure *i*-butane is produced by this process via a gas phase reaction in the presence of hydrogen and organic chlorides. A flowchart is given in Fig. 3.9.

The second important process is the C$_4$-isomerisation of BP Chemicals [3.34], in which extremely pure *i*-butane is produced from *n*-butane using a solid bed catalyst containing platinum. *i*-Butane is continuously removed from the circulating stream in a column. The yields are virtually quantitative. Figure 3.10 illustrates the main features of this process.

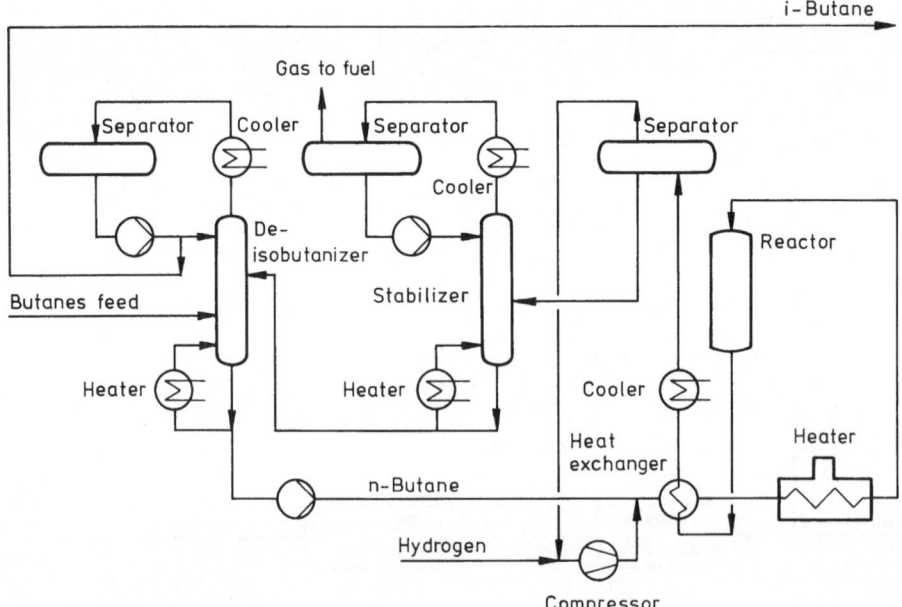

Fig. 3.9. The UOP Butamer process [3.33]

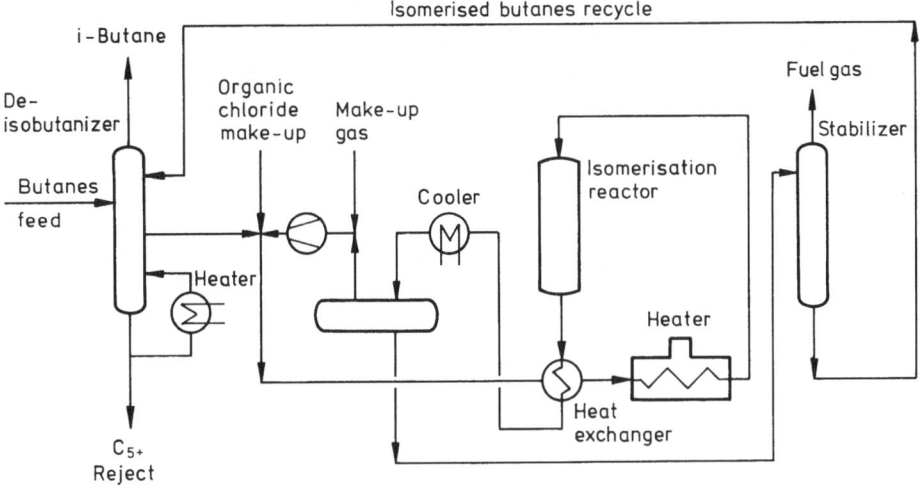

Fig. 3.10. The BP C$_4$-isomerisation [3.34]

3.2.2 Dehydrogenation

The dehydrogenation of *n*-butane gives a mixture of butenes which can be further dehydrogenated or oxydehydrogenated to butadiene:

$$CH_3-CH_2-CH_2-CH_3 \xrightarrow{-H_2} CH_2=CH-CH_2-CH_3 + CH_3-CH=CH-CH_3$$
n-Butane $\qquad\qquad$ 1-Butene $\qquad$ cis+trans 2-Butene

$$CH_2=CH-CH_2-CH_3 + CH_3-CH=CH-CH_3 (+1/2 O_2) \xrightarrow{-H_2(-H_2O)} 2CH_2=CH-CH=CH_2$$

Both reaction steps are endothermic and each requires an energy supply of about 2100 to 2300 kJ/kg butane used. The reactions proceed best at high temperatures (600 to 700 °C) and low pressure on catalysts which suppress the kinetically favoured crack reactions.

i-Butane can be similarly dehydrogenated to *i*-butene. Oxidation of the *i*-butene can yield methacrolein:

$$\begin{array}{c} CH_3 \\ CH_3 \end{array}\!\!>\!CH-CH_3 \xrightarrow{-H_2} \begin{array}{c} CH_3 \\ CH_3 \end{array}\!\!>\!C=CH_2 \xrightarrow{+O_2} CH_2=\underset{\underset{CH_3}{|}}{C}-CHO + H_2O$$
i-Butane $\qquad\qquad$ i-Butene

The dehydrogenation of *n*-butane plays a significant part only in Eastern Europe at present, mainly for producing butadiene. After the second world war it soon became unimportant in Western Europe. It is practised only occasionally in the existing plants in North America on account of the different raw materials situation in petrochemistry.

Structural changes of refineries in the USA have led to increased amounts of available refinery butenes, not all of which are used to prepare alkylate gasoline. Consequently, since the middle of the 1960s oxydehydrogenation processes of *n*-butene have been operated as a new process generation which displays technical advantages over the *n*-butane dehydrogenation processes. Oxydehydrogenation of butanes would not make sense because at the high temperature level too many side products would be formed. A process of this kind from Shell, described in the literature, which made use of elementary iodine as a hydrogen acceptor, was consequently never capable of being exploited on the commercial scale [3.35]. Sometimes the dehydrogenation of *i*-butane has been performed, but only in Eastern Europe, mainly to make isoprene from the *i*-butene and to reduce dependence on imports of natural rubber which consumes valuable foreign currency.

In the meantime there are signs of a shortage of *i*-butene in the western world, brought about by the MTBE boom. This additional demand could be covered by dehydrogenation and isomerisation of LPG butanes. An increasing availability of butanes as natural gas condensate or associated gas from crude oil (largely flared at present) is arousing increased interest in new dehydrogenation plants in the oil producing regions in order to produce MTBE from the butane. Reference is made here to new developments towards more flexible procedures alongside the classical dehydrogenation of *n*-butanes. The oxydehydrogenation of butenes is described in Sect. 3.3.3.2.

The older processes for dehydrogenating *n*-butane are essentially based on development work by Houdry and Phillips Petroleum. The dominating procedure, called the Houdry process but more exactly termed the Catadiene process, was employed in most of the plants based on butane gas for butadiene production [3.36]. In this one-step process the gas is directly dehydrogenated on an Al/Cr-oxide catalyst to a mixture of *n*-butenes and butadiene. A disadvantage of the process is the rapid deactivation of the catalyst through coke deposit on its surface. The cyclical use of at least three reactors in parallel is thus obligatory. After 5 to 15 min reaction time in each, the rectors must be regenerated, which involves burning off the coke residue, furnishing heat for the next reaction period. Unreacted butane is recycled together with the butenes. The yield of butadiene is about 55%. Saturated and unsaturated C$_1$-C$_3$-hydrocarbons plus small amounts of carbon oxides are yielded as co-products. The demonstration plant, erected in 1944 in El Segundo, CA (18,000 t/a butadiene), was followed at the end of the 1950s by plants under licence — Texas Butadiene (later Petro-Tex, 135,000 t/a); Firestone (100,000 t/a) in Orange, TX; Odessa Butadiene (later El Paso, 50,000 t/a); Sinclair (later Arco, 125,000 t/a) in Channelview, TX; and Hüls (40,000 t/a) in Marl. Up to now 17 Catadiene plants have been erected throughout the world. The last Catadiene plant to be built was in the USSR in 1985. It has a capacity of 180,000 t/a butadiene. Two of the western world Catadiene plants are still kept in a state ready for operation, although they have scarcely been used since the beginning of the 1980s. Firestone and El Paso were running them up to the middle of 1982.

In the older, two-step Phillips butadiene process *n*-butane is first dehydrogenated according to the above equation, on Cr/Na-oxide catalysts, yielding the butenes which are separated from the residual butane which has not reacted [3.37]. For the second step the butenes are mixed with superheated steam and converted to butadiene in an externally heated tubular reactor containing a Fe/K/Al-oxide catalyst. Unreacted *n*-butane is returned to the first step. This procedure has the advantages of longer cycling times (about 1 hour) and higher butadiene yields (65%). Phillips was running a plant in Borger, TX since 1944, which produced a maximum of 65,000 t/a butadiene.

Two processes have been used to prepare butenes from the butane fraction from coal hydrogenation installations: the Leuna process in the German Leuna-Werk of the I G and the UOP process at the British ICI factory at Billingham. In addition to the catalyst problems mentioned, these older procedures suffer above all from the low concentrations of product in the reaction gases. The butenes and butadiene have to be separated from these gases by means of extractive distillation; the separation of butadiene in low concentrations of 15 to 18% as in the Catadiene process, is more difficult than when high concentrations are available, as for example up to 50% in steam cracker fractions.

In recent years three organisations have been striving to develop more flexible processes which could become of economic importance. They offer advantages as an alternative to butane cracking especially when the cracking products ethylene and propene have limited outlets at the particular site. This is true when refineries want to obtain butenes from butane-rich waste gases for purposes of alkylation or MTBE.

Table 3.3 shows the better yields of butenes obtained by dehydrogenation compared to cracking. Obviously i-butene can be obtained in about 90% yield from i-butane with recycling and isomerisation of the n-butanes/butenes. Additional advantages over cracking are the lower reaction temperatures (550 to 650 °C compared to 800 to 1000 °C), the lower capital requirements and the formation of fewer low-value co-products such as methane and ethane.

Licenses have been offered for the following new dehydrogenation processes:

The Catofin process of Air Products and Chemicals, Allentown, PA permits the dehydrogenation of LPG to propene-butene mixtures or of the components propane or butanes to the respective olefines [3.39, 3.40].

A flow chart of the process is given in Fig. 3.11. It is a further development of the classical Houdry process. Texas Petrochemicals (formerly Petro-Tex) commenced operation a short time ago in Houston, TX of a 550,000 t/a plant for MTBE, based on i-butane. Its capacity is now being doubled, for which parts of the old Houdry plant have been revamped. This reconstruction of a Houdry unit (350,000 t/a i-butene) cost about $ 20 million. A new Catofin plant of this size would have cost about $ 80 million [3.40]. In spring 1988 another i-butene dehydrogenation plant of Saudi European Petrochemical is coming into operation at Al Jubail in Saudi Arabia. The national Petroleos Mexicanos (Pemex) is building a plant with a capacity up to 350,000 t/a propene from propane by the same process in Morelos.

Table 3.3. Comparative yields from butanes by catalytic dehydrogenation and by thermal cracking [3.38]

Raw material Product/yield in wt. %	i-Butane		n-Butane	
	Catalytic dehydrogenation	Thermal cracking	Catalytic dehydrogenation	Thermal cracking
Hydrogen	3.0	2.0	3	2
Ethylene	–	20.0	2	35
Propene	3.0	30.0	2	28
i-Butene	80.0	42.0	2	1
n-Butenes	9.0	1.0	79	19
Butadiene	0.3	1.0	2	4
i-Butane	–	–	2	–
n-Butane	1.7	1.0	–	–
Remainder	3.0	3.0	8	11

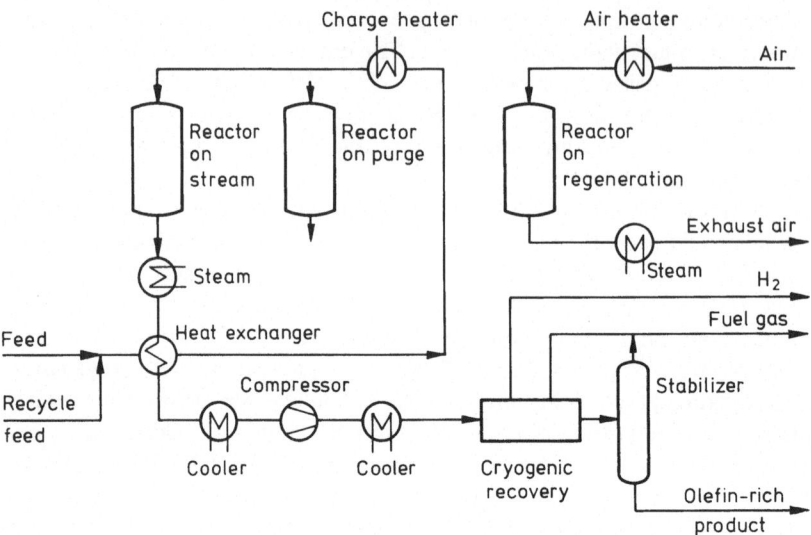

Fig. 3.11. Catofin process of Air Products and Chemicals [3.42]

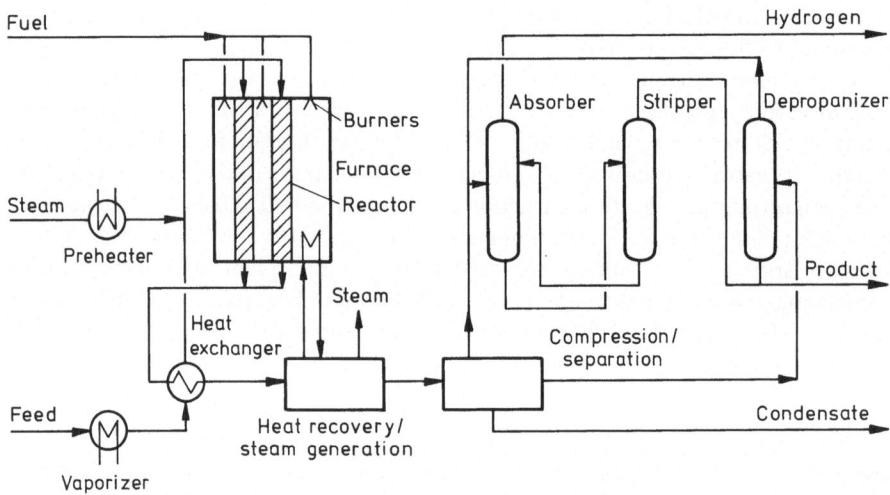

Fig. 3.12. Phillips Star process [3.42]

The first stage of the older Phillips process has likewise been improved. It is now available as a one-step version for deriving butenes from butanes under the name of Star (Steam Active Reforming Process) [3.43]. The flow chart in Fig. 3.12 shows the two-section tubular reactor. The tubes can be alternately regenerated after about eight hours cycle time. The process has been demonstrated on the semi-commercial scale.

UOP is also granting licenses for a modern, flexible LPG-dehydrogenation process [3.44]. This process, offered under the name of Oleflex, operates with a fluid bed catalyst. This is a technology which has been applied successfully in Platforming (catalytic reforming) in refineries [3.38]. Figure 3.13 shows a flowchart of the process.

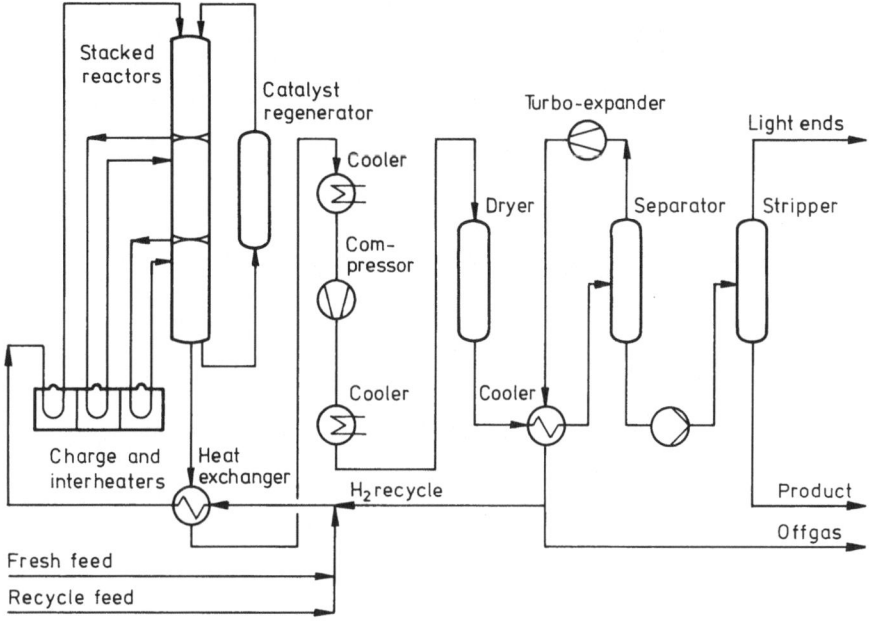

Fig. 3.13. UOP Oleflex process [3.42]

The catalyst circulation can be varied to a time interval of three to fourteen days. The first commercial Oleflex unit will go on stream mid-1989 in Thailand and will have a capacity of 105,000 t/a propene made from propane.

3.2.3 Oxidation

Formerly only *n*-butane was subjected to oxidation and only later were processes developed for *i*-butane oxidation. The procedures for oxidising *n*-butane can be divided into those in the gas phase and those in the liquid phase. The older gas phase oxidations yielded an immense variety of products, including C_1- and C_2-acids, aldehydes, ketones, alcohols and acetates. The laborious separation procedures and meagre yield of the products of principal interest made gas phase oxidation appear uneconomical in comparison with oxidation in the liquid phase.

Liquid phase oxidations usually have acetic acid as the main product. The plants hitherto built for this purpose are listed in Table 3.4 [3.45].

By the end of 1987 there was only one *n*-butane oxidation plant in use in the world — that of Celanese in Pampa, TX but it was destroyed by a big fire. Its capacity was 263,000 t/a acetic acid, plus co-products. During normal operation, up to 40,000 t/a MEK, 23,000 t/a formic acid, 18,000 t/a ethyl acetate and 2000 t/a ethanol were produced. Large deposits of wet natural gas sometimes enabled further plants to be run there, e. g. that of Celanese in Bishop and that of UCC in Brownsville. In the FRG Hüls also ran a butane oxidation of this sort, for 24,000 t/a acetic acid, from 1961 — however, it was soon closed down as a result of difficulties with raw materials supply

Table 3.4. World liquid phase oxidation processes for *n*-butane [3.45]

Country	Company	Site	Acetic acid capacity (1000 t/a)	Production start-up
Canada	Celanese	Edmonton, AL	64	1956
USA	Celanese	Bishop, TX	90	n. a.
	Celanese	Pampa, TX	263	1952
	UCC	Brownsville, TX	295	1961
FRG	Hüls	Marl	24	1961
Netherlands	AKZO-Zout*	Rotterdam	110	1965
USSR	state-owned refinery	Moscow	n. a.	1962

* formerly Konam (joint venture Celanese and Zout-Ketjen)

and corrosion [3.46]. Another plant, of AKZO-Zout in the Netherlands, was closed down because of corrosion damage.

Maleic anhydride (MA), formerly almost exclusively obtained by oxidation of benzene, has become available, since the 1970's via catalytic oxidation of *n*-butane in the gas phase [3.47].

$$CH_3-CH_2-CH_2-CH_3 \quad + \quad 7/2\ O_2 \quad \longrightarrow \quad \underset{CH}{\overset{CH}{\|}}\underset{\diagdown C \diagdown\diagup}{\overset{\diagup C \diagup}{}}O \quad + \quad 4\,H_2O$$

This oxidation of *n*-butane is carried out at 440 to 400 °C, usually on a modified V/P- oxide-catalyst, for which there is comprehensive patent literature from many companies active in the petrochemical field (BASF, Halcon, Monsanto, Petro-Tex, Amoco, Standard Oil, Mitsubishi etc.). The yields of MA lie between 45 and 60%. Plants at present using as the starting material benzene can be modified and adapted to convert *n*-butane. The catalyst and reaction parameters must, however, be adjusted accordingly. On account of a smaller space-time-yield the MA-capacity of a plant converted to the use of butane is only 60 to 70% at full capacity utilization compared to that on the basis of benzene [3.48].

Technical progress has also been achieved in MA-recovery from the reaction gas. Hitherto it was possible to condense directly only half of the MA in the butane production method (60% in the benzene oxidation). The other half had to be removed in solution in water as dilute maleic acid, from which MA is derived by dehydration with *o*-xylene at high temperature. Problems of corrosion and waste water disposal associated with this second stage can be avoided by using an organic absorption agent as in the Alma process of Alusuisse Italia/Lummus Crest [3.49]. A semi-commercial plant (3000 t/a) with a fluid bed reactor began test production in Scanzorosciate in 1984. The catalyst and process parameters were optimised [3.50]. The costs for the catalyst and absorption agent have been kept confidential. Alusuisse has decided to build its own large scale plant and is interested in granting licenses [3.51]. The first commercial Alma plant is due to commence production at the end of 1988 at Shin-Daikyowa Petrochemical, Yokkaichi [3.52]. A flow chart of the process is shown in Fig. 3.14.

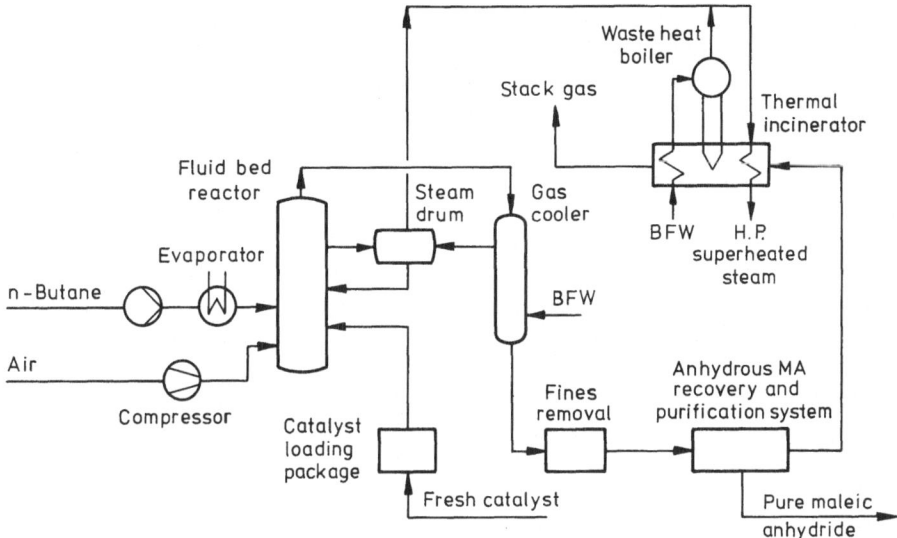

Fig. 3.14. Production of MA according to the Alma process [3.50]

The Alma process with only slight modifications is suitable also for producing fumaric acid. This contains one molecule more water than MA. The energy-consuming dehydration stage needed for production MA is not necessary for fumaric acid because it is yielded directly through isomerisation of maleic acid [3.53]. The lower-priced raw material basis and the more severe requirements for protection of the environment in connection with benzene emission have led, in the USA, to wide-spread substitution of benzene by butane as starting material for preparing MA. This has probably come to an end now [3.54]. In the FRG only benzene and *n*-butenes serve as a basis for MA (see Sect. 3.3.3.4). It is obtained also as a co-product of phthalic anhydride (PA) production.

The oxidation of *i*-butane to *tert*-butyl hydroperoxide (TBHP) and the following reaction of this intermediate with propene to give TBA and propene oxide have been already described, in Sect. 2.3.2. This process, which is only being operated by Arco in the Netherlands, USA and, from 1988, also in France, makes available a number of interesting products. TBHP is a valuable polymerisation initiator for plastics such as low density polyethylene, polystyrene, PVC and ABS. Since the middle of 1979 about 3000 t/a of it have been obtained from the production in the Netherlands and sold as a 70% solution [3.55]. This Arco process is not yet licensed to other companies.

According to developments at Mitsubishi [3.56] and Arco/Halcon [3.55], TBA can be oxidised in a two-stage process on a solid bed catalyst to methacrylic acid and finally esterified to methyl methacrylate (MMA):

$$CH_3-\underset{\underset{CH_3}{|}}{\overset{\overset{CH_3}{|}}{C}}-OH \xrightarrow[-H_2O]{+O_2} CH_2=\underset{\underset{}{\overset{CH_3}{|}}}{C}-CHO \xrightarrow{+1/2 O_2} CH_2=\underset{\overset{CH_3}{|}}{C}-COOH \xrightarrow[-H_2O]{+CH_3OH} CH_2=\underset{\overset{CH_3}{|}}{C}-COOCH_3$$

TBA   Methacrolein   Methacrylic acid   Methyl
(or *i*-butene)   methacrylate (MMA)

Halcon has recently announced a variant of this process, which, as a synergistic integration of an i-butane dehydrogenation and an i-butene oxidation to methacrolein, is given good economic chances as a future production method for MMA. The following provisional specific consumption data were announced per t MMA [3.57] : 0.92 t i-butane (95 wt. %); 1.44 t oxygen; 0.34 t methanol. The capital outlay, not counting the butane isomerisation, is of the same order of magnitude as that for a plant using the acetone cyanohydrin (ACH)process at the same capacity. No information was given about product quality specifications.

The reaction products from both oxidation stages are processed further in a single installation. According to patents of Nippon Kayaku the process can be simplified further by direct esterification to MMA as early as the second oxidation stage, on Pd/ P-oxide catalysts [3.58].

However, the high cost of methanol recovery and poor conversion still make this route economically uninteresting, as was proved by similar studies by Asahi [3.57].

Methacrolein and methacrylonitrile are, in principle, available also through catalytic oxidation or ammonoxidation of i-butane. The older technologies, are, however, uneconomical, as past work of ICI and Power-Gas Comp. revealed. Using V/Sb-containing oxide catalysts, molar yields of 2.1% at space-time yields of about 500 g methacrolein or methacrylonitrile per l and h were attainable in a cycle process [3.59]. The feedstock mixture of i-butane and air had each time a high butane content (80%).

Among other processes Rohm and Haas have developed a single-stage oxidation of i-butane to methacrylic acid. Selectivities were quoted for methacrylic acid of up to 50% and for the co-product, methacrolein, of up to 20%. High concentrations of i-butane and large reactors were needed, yet conversion rates were poor [3.57].

3.2.4 Alkylation

An alkylate is obtained by reacting i-butane with C_3-C_5-olefines in the liquid phase with concentrated sulfuric acid or anhydrous hydrofluoric acid as catalyst. The product consists essentially of branched C_5-C_{10}-paraffins and is used as a gasoline component of high octane number. Alkylate gasolines have a high lead sensitivity, i. e. addition of only very little lead alkyls increases the octane number appreciably. As a result of this effect alkylates are, in contrast, less suitable as components of lead-free gasoline.

Refinery B-B is most suitable as raw material for alkylate manufacture because of its otherwise limited outlets. The C_4-cut from steamcrackers seems to be too valuable as a raw material for making alkylate on account of its relatively high content of i-butene for which there is an ever-increasing demand as an intermediate chemical. The development of the Dimersol G process of IFP, in which propene is dimerised catalytically in the liquid phase with 90% yield to i-hexenes which have high research octane number, has led to a more limited use of propene together with i-butane in alkylation processes [3.60].

As a consequence of the refinery structure, the alkylation is carried out above all in North America. Increased alkylate gasoline capacities are conceivable in Europe on account of the progressive structural changes of European refineries and the exten-

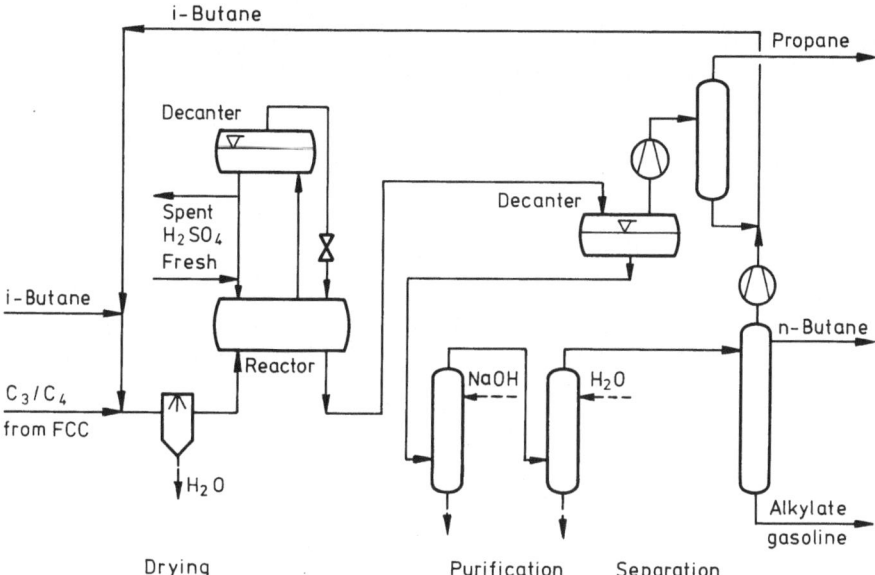

Fig. 3.15. Stratco alkylation using sulfuric acid [3.60]

sion of catalytic cracking. The alkylate gasoline capacity in the western world was about 41 million t/a in 1983, of which about 75% were butene alkylate. The share of the sulfuric and hydrofluoric acid processes was in each case about one half of the total capacity [3.61].

The Kellogg and Stratco processes are the most important alkylations with sulfuric acid (Fig. 3.15). Plants based on the Stratco process alone have a worldwide capacity of 50,000 t/day [3.62]. Alkylation using sulfuric acid is not carried out in the FRG.

i-Butane and the olefines in the ratio of between 6 and 10:1 are reacted at 4 to 5 °C in the presence of concentrated sulfuric acid in the Stratco reactor. The ratio of acid to hydrocarbon is about 1:1. Losses of sulfuric acid are compensated by introducing fresh acid (98%). It has been possible to reduce acid consumption to 75 to 80 kg/t alkylate. Formation of acid tar is almost completely suppressed through the high excess of *i*-butane. The reaction heat is removed with circulated *i*-butane [3.62].

In the Kellogg process the conversion takes place in a cascade reactor. The sulfuric acid taken off as side stream can be recovered by the SARP process. This involves the reaction of sulfuric acid with propene to dipropyl sulfate, which is extracted from the acid tar with *i*-butane. This extract can then be returned to the alkylation reactor. Acid tar is pyrolytically split into coke and sulfur oxides. The latter can be reprocessed to sulfuric acid [3.61].

UOP and Phillips Petroleum license alkylation processes with hydrofluoric acid. The sole existing German plant of Oberrheinische Mineralölwerke [OMW, 42% Deutsche Texaco, 33% Veba Oel (Aral), 25% Conoco (Jet)] in Karlsruhe has been in operation since 1979 equipped for the UOP process and has a capacity of 300,000 t/a butene alkylate. A flow chart of the UOP process is given in Fig. 3.16.

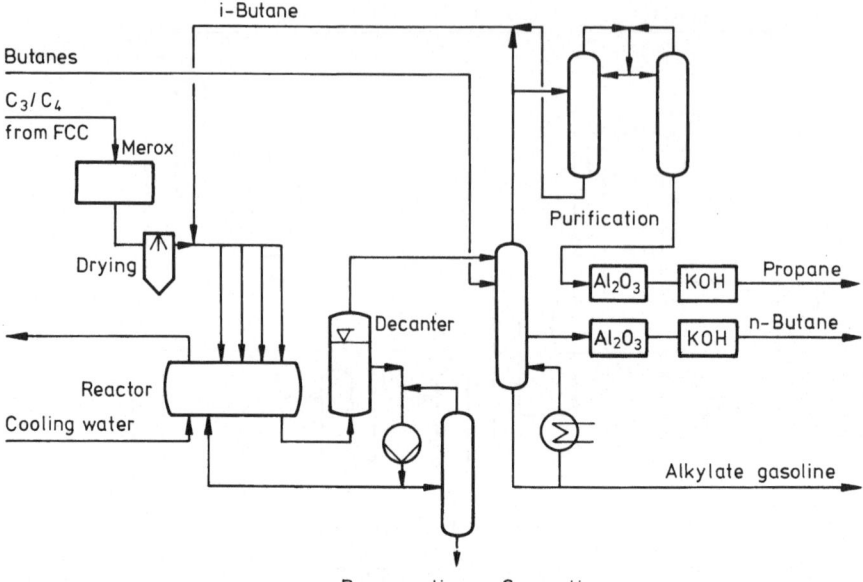

Fig. 3.16. UOP alkylation using hydrofluoric acid [3.60]

The alkylation is performed with anhydrous hydrofluoric acid at reaction temperatures of 10 to 36 °C. During the reaction the ratio of i-butane to olefines is between 3 and 10:1, that of hydrocarbon to acid is 2:1. The hydrofluoric acid is circulated and regenerated each time in a side stream. About 0.15 to 0.20 kg acid is lost per ton alkylate [3.62]. The hydrofluoric acid is extremely corrosive and hence a large part of the plant has to be constructed of Monel metal. A comparison of the competitive aspects of the sulfuric acid and hydrofluoric acid processes is described in Sect. 6.1.5.1. About 0.45 t butenes and 0.60 to 0.63 t i-butane are required to produce 1 t butene-alkylate.

Research with solid catalysts for the i-butane/olefine alkylation is presently in progress. Several research teams have succeeded in carrying out alkylation using wide-pore zeolites (faujasites) or organic ion exchangers (e. g. Nafion). An obstacle to technical application is still, however, the rapid deactivation of the catalysts caused by deposits in the pores [3.61].

3.2.5 Chlorination

The thermal treatment of butane with chlorine gas yields a wide variety of chlorobutanes, the relative proportions of which are determined chiefly by the chlorine/butane ratio. Approximate amounts are 35 to 50% 2-chlorobutane, 25 to 35% 1-chlorobutane, 5 to 15% 1,3-dichlorobutane, and smaller amounts of the other dichlorobutanes. Chlorination of butane is not carried out technically at present because most of the co-products cannot be utilized profitably. It is better to derive chlorobutanes by esterification of the butanols with hydrochloric acid. Various syntheses are possible and are stated below [3.64].

n-Butyl chloride (1-chlorobutane) is obtained by esterifying n-butanol with hydrogen chloride or aqueous hydrochloric acid, if necessary in the presence of zinc chloride:

$$CH_3-CH_2-CH_2-CH_2OH \; + \; HCl \; \longrightarrow \; CH_3-CH_2-CH_2-CH_2Cl \; + \; H_2O$$

The yield is about 73%. n-Butyl chloride is used as a solvent for oils, fats and waxes, for removing water from fats and acids, and in Friedel-Crafts reactions. It is also a base chemical for bis-tributyltin oxide, $(C_4H_9)_3SnOSn(C_4H_9)_3$, which is used as an anti-fouling additive for sea-water paints and as a fungicide.

i-Butyl chloride (1-chloro-2-methylpropane) is generally prepared by esterifying i-butanol with HCl:

$$\begin{array}{c} CH_3 \\ CH_3 \end{array}\!\!\!\!CH-CH_2OH \; + \; HCl \; \longrightarrow \; \begin{array}{c} CH_3 \\ CH_3 \end{array}\!\!\!\!CH-CH_2Cl \; + \; H_2O$$

It can be used in Friedel-Crafts reactions but only small amounts are required.

sec-Butyl chloride (2-chlorobutane) is prepared by esterifying sec-butanol with HCl:

$$CH_3-CH_2-CHOH-CH_3 \; + \; HCl \; \longrightarrow \; CH_3-CH_2-CHCl-CH_3 \; + \; H_2O$$

The product is used in Friedel-Crafts syntheses, e. g. with benzene and $AlCl_3$ to yield sec-butylbenzene, with benzene and Al/Hg to give tert-butylbenzene, and also for sec-butyl-1-naphthalene.

tert-Butyl chloride (2-chloro-2-methylpropane) is also almost always prepared by esterifying the appropriate alcohol, namely, tert-butanol, with HCl:

$$\begin{array}{c} CH_3 \\ CH_3 \end{array}\!\!\!\!COH-CH_3 \; + \; HCl \; \longrightarrow \; \begin{array}{c} CH_3 \\ CH_3 \end{array}\!\!\!\!CCl-CH_3 \; + \; H_2O$$

It can also be prepared by addition of HCl to i-butene in the gas phase at temperatures below 100 °C on metal oxides and chlorides, e. g. $AlCl_3$, or also at low temperatures:

$$\begin{array}{c} CH_3 \\ CH_3 \end{array}\!\!\!\!C{=}CH_2 \; + \; HCl \; \longrightarrow \; \begin{array}{c} CH_3 \\ CH_3 \end{array}\!\!\!\!CCl-CH_3$$

In both processes consumption of HCl is advantageous but there is little demand for the product. This is also used in Friedel-Crafts reactions, e. g. for preparing tert-butylbenzene and tert-butylphenol, as well as 4-chloro-2,2-dimethylbutene (neohexyl chloride) as a base chemical for synthetic perfumes.

1,4-Dichlorobutane can likewise be obtained by esterification of 1,4-butanediol with HCl:

$$CH_2OH-CH_2-CH_2-CH_2OH \; + \; 2HCl \; \longrightarrow \; CH_2Cl-CH_2-CH_2-CH_2Cl \; + \; 2H_2O$$

1,4-Dichlorobutane was formerly an important intermediate in the production of synthetic fibres (Nylon). It can be converted to adiponitrile with HCN or sodium cyanide and this can in turn be hydrogenated to hexamethylenediamine. Hydrolysis of the dinitrile gives adipic acid which reacts with the hexamethylenediamine to yield Nylon salt, a base material for making polyamide 6.6. At present these processes are of no practical importance for Nylon production. However, butadiene seems to be becoming an increasingly more favourable raw material in this field (see Sect. 3.4.3).

3.2.6 Cyclisation

Aromatics can be produced from LPG by an analogous application of the principles of the reforming processes which have been customary for a long time with C_5-C_{10}-fractions in most refineries. BP and UOP together have developed the Cyclar process in which about two thirds of the LPG or refinery gases are converted to aromatics. A flow chart of the process is given in Fig. 3.17.

The pre-heated raw material and the zeolite catalyst are fed continuously to the reactor head. Low-boiling components are separated from the reactor gas. Technically valuable hydrogen and a fuel gas (methane/ethane) can be separated from the reactor gas, rendering the process (including the compressor drive) self-sufficient in energy consumption. The catalyst is relatively resistant to small amounts of impurities in the raw material, for example sulfur, water, CO_2 and oxygenates. However, the quality requirements for hydrogen mean that sulfur has to be removed from the raw materials.

The capital outlay for the process has only been announced so far for the process using propane. It amounts, in this case, to 46.6 million US \$ [3. 65] for a propane throughput of 333,000 t/a, corresponding to a production capacity of 212,000 t/a C_{6+} or 148,000 t/a C_{7+} gasoline rich in aromatics [3.64]. Often, the benzene must be removed because of environmental restrictions. No information was provided about specific consumption figures. The gasoline properties obtained are:

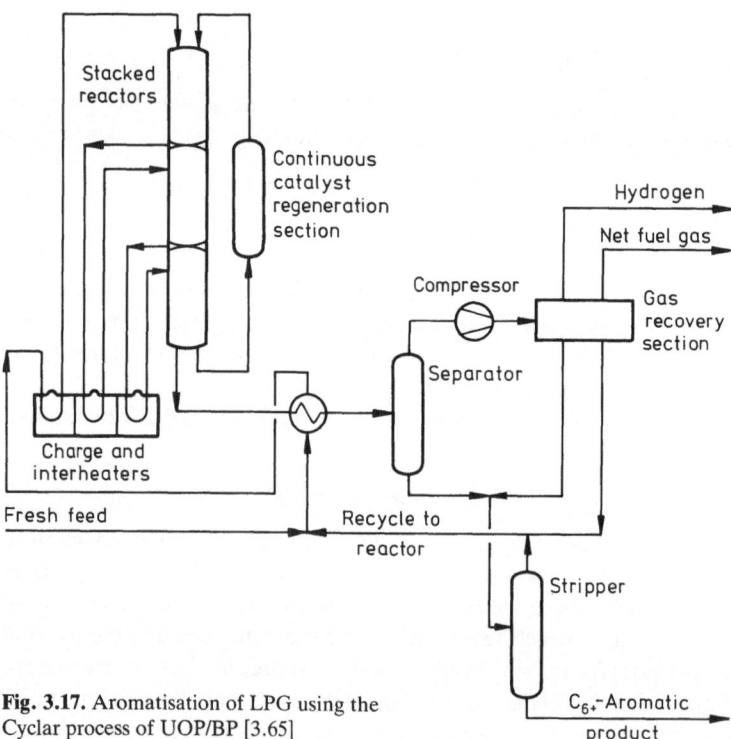

Fig. 3.17. Aromatisation of LPG using the Cyclar process of UOP/BP [3.65]

	C_{6+}-fraction	C_{7+}-fraction
RON	112	113
MON	100	102
Vap. pressure (Reid) in bar	0.112	0.049

Table 3.5. Yields in wt. % from the Cyclar process [3.65]

Fraction	Products/Raw materials	Propane	Butane
C_6	Benzene	19.3	18.0
C_7	Toluene	26.6	28.9
C_8	Ethylbenzene, xylenes	11.1	13.4
C_{9+}	Higher aromatics	6.6	7.2
	Hydrogen (95%) $\varsigma = 0.09$ kg/m^3	6.0	5.5
	Refinery gas	30.4	27.0

The first Cyclar demonstration plant for the production of 20,000 t/a gasoline, should go into operation at the end of 1988 at BP in Grangemouth.

3.2.7 Other Applications

A large part of the butanes is directly blended with the gasoline for Otto engines, in order to enhance the volatility.

The demand for gasoline fuel fluctuates seasonally, reaching its peak in the summer months. In warmer climates the demand likewise depends on the season but there demand is lower in summer because the gasoline mixture then evaporates more rapidly than in the cooler winter periods. According to DIN 51600 the vapour pressure of all gasoline fuels for Otto engines in the FRG may be 0.9 bar in winter and 0.7 bar in summer (test conditions according to Reid, DIN 51754). This means that in the summer months about 6% by wt. of butanes can be mixed with the fuel; the average for the whole year can be estimated as 7% [3.66].

As a result of measures to protect the ozone layer of the stratosphere one has to expect that in the USA legal requirements for a reduction of the gasoline vapour pressure in the summer months will be implemented. The amount of butanes blended in the American Gulf coast refineries is to be lowered from the present 5.52% by volume to only 1% after 1990 [3.67]. A uniform regulation in the USA will not be easy to enforce because of the differing climatic zones.

Ways of utilizing the future butane raw material potential to the maximum advantage are being investigated at present. Increased use as steam cracker feed may be accompanied by intensified utilization in the competitive processes of production of MTBE, alkylate gasoline and aromatics [3.68].

The use of propane/butane mixtures (LPG) as an alternative fuel for motor vehicles, and hence the improved utilization of these resources, was discussed intensively during the oil crises. LPG has a distinctly lower volume-specific calorific value than gasoline, so that from 15 to 33% by volume extra consumption, depending on the mode of driving, is experienced. The higher transport and storage costs and the inevitable increase in weight of the vehicles as a result of protecting the gas tank against accidental damage have restricted the use of LPG as a motor fuel. Of the industrialized nations, the Netherlands and Italy have, relatively, the most LPG stations and LPG-powered motor vehicles [3.69].

Production of synthesis gas from refinery gas may sometimes be advisable, e. g. to compensate for the hydrogen consumption of a hydrocracker. Steam reforming yields synthesis gas according to the equation:

$$CH_3-CH_2-CH_2-CH_3 \; + \; 4\,H_2O \longrightarrow 4\,CO \; + \; 9\,H_2$$

Considerable amounts of butane, accompanying methane in natural gas, are used for producing synthesis gas, CO and H_2, leading to products such as methanol, ammonia, oxo-products, etc.

In the FRG, Hüls and URBK together supply about 5000 to 7000 t/a i-butane as a propellant for aerosols. In addition, i-butane is a solvent used in the Phillips HDPE process (approx. 1% of i-butane related to HDPE).

A technical synthesis of thiophene depends on the reaction of butane with sulfur or carbon disulfide [3.70].

Residual C_4-hydrocarbons which find use neither in the chemical industry nor in refineries, i. e. principally butanes from C_4-cuts or refinery gases, are sometimes fed as raw material into a steam cracker for producing ethylene or an electric arc for producing acetylene [3.71]. If there are no such plants available in the vicinity (transport costs) only their use as heating gas remains. Butane gas burns cleanly and is thus excellent for heating steam boilers and reactors. The relatively high volume-specific calorific value is advantageous in relation to the storage costs. Small pressurised cylinders of butane are commercially available (e. g. for camping).

3.3 Processes of Butene Conversion

The presentation below of the treatment of butenes is subdivided according to the available raw materials in the further utilization of refinery B-B and raffinate I. It shows which C_4-chemistry processes, after stepwise separation of the raw materials, can be integrated into production schemes at the present state of technical progress. The separation of the raw material is governed by the scheme in Fig. 3.3 (see Sect. 3.1.2.1). This type of subdivision follows the needs of process planning so as to ascertain the best technical ways of upgrading products in a refinery or a chemical factory.

First, the technical processing of the refinery butenes is treated. The preparation of fuel components with high octane numbers is emphasized. Ways of processing chemical butenes are then discussed in the sequence of their degrees of purity, beginning with the raffinates and concluding with the pure components.

Since the butanes are inert during most butene processes they need not be separated from the product streams. Sometimes they may even exert synergistic effects.

3.3.1 Components of Gasoline Fuels

Fuel additives, as active ingredients (e. g. catalysts, inhibitors) are added to gasoline only in small amounts. In contrast, motor fuel components have, because of their chemical compositions, good combustion properties which enables them to be used as ordinary fuels.

The following fuel components can be derived from butenes:

— polymer gasoline
— alkylate gasoline
— *tert*-butanol (TBA)
— methyl *tert*-butyl ether (MTBE)
— *sec*-butanol (SBA)

Since the production of alkylate gasoline requires *i*-butane as well as the butenes, the processes have been treated previously — (see Sect. 3.2.4). The production of TBA and SBA from butene-containing fractions has also been described previously since the procedures serve predominantly for raffinate separation (see Sects. 3.1.2.2 and 3.1.2.4). Hence only the production of polymer gasoline and MTBE is discussed here.

3.3.1.1 Polymer Gasoline

Polymer gasoline is a branched C_6-C_{12}-oligomerisate, prepared from mixtures containing propene and butene. The composition of the mixtures varies over a wide range but it can be assumed that, on the average, the olefinic part consists of about 60% by wt. of propene and 40% by wt. of butenes. These last-named contain about two thirds *n*-butenes and one third *i*-butene and under the usual operating conditions, they are converted almost entirely into polymer gasoline. For a long time the chief advantage was the conversion in this way of the propene and butene constitutents of FCC-cuts into gasoline components of relatively high research octane number, whereas they diminish the quality of alkylate gasoline when present.

At first sulfuric acid was the catalyst used but it has since been displaced in this field. The mixtures reacted at 90% conversion degree in reactors at temperatures up to 150 °C and pressures up to 50 bar. The ratio of the dimers to trimers thereby formed is 4 : 1. Currently, the most used polymerisation procedure in refineries, that of UOP, employs solid bed catalysts (phosphoric acid on asbestos or kieselguhr carriers) which are distributed over five compartments of a reactor. The heat of reaction is removed with a circulating propane/butene stream [3.72].

In the Kellogg process, polymerisation is carried out in stages, in tubular reactors containing copper pyrophosphate as catalyst [3.72].

Fixed bed reactors are used in the process of California Research. They contain quartz sand which is soaked with the catalyst, phosphoric acid.

In the Selectopol process, licensed by IFP, *i*-butene from raffinate I or refinery B-B is selectively polymerised on an acid catalyst in a multiple solid bed column. Unreacted raffinate is circulated as a coolant [3.73]. This newest polymer gasoline process has not yet become commercially established because it competes for raw material with the more rewarding production of MTBE. Figure 3.18 shows a flow chart of the process.

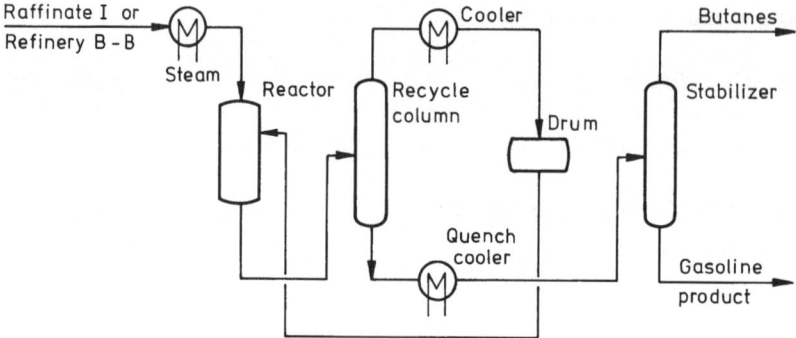

Fig. 3.18. IFP Selectopol process for production of polymer gasoline [3.73]

The dimer and trimer products are then always washed with alkali and water and hydrogenated to i-octane or i-dodecane on a nickel catalyst at temperatures of 180 to 200 °C and pressures up to 200 bar. Extensive isomerisation takes place during this step. n-Butenes which have not reacted have to be co-hydrogenated in the IFP process (Hydropol process). After separation the butanes formed are, for example, fed to an ethylene plant [3.73].

Existing refinery capacities in the FRG for producing polymer gasoline are: Deutsche Shell, Hamburg (35,000 t/a); Deutsche Texaco, Heide (25,000 t/a); and URBK, Wesseling (75,000 t/a). It is known from some refineries producing polymer gasoline that they separate the heptenes before hydrogenation, in order to use them for the manufacture of plasticisers. Esso is the world leader in this market. The British Esso refinery in Fawley, which has a long-term contract to supply ICI, has the largest capacity of 200,000 t/a heptenes, of which approx. 110,000 t/a are n-heptenes. CFR in France, the sole European competitor, has a capacity of 10,000 t/a. Heptene producers in North America are: Amoco in Yorktown, VA (10,000 t/a), Exxon in Baton Rouge, LA (20,000 t/a), Getty Oil in Delaware City, DE (20,000 t/a), and Imperial Oil Canada, which belongs to Esso (20,000 t/a).

3.3.1.2 Methyl *tert*-Butyl Ether (MTBE)

The catalytical conversion of i-butene with methanol gives methyl *tert*-butyl ether (MTBE) according to the equation:

$$\begin{array}{c}CH_3\\ \diagdown\\ C{=}CH_2\\ \diagup\\ CH_3\end{array} + CH_3OH \longrightarrow \begin{array}{c}CH_3\\ \diagdown\\ CH_3{-}C{-}O{-}CH_3\\ \diagup\\ CH_3\end{array}$$

The product MTBE is the result of intensive R and D efforts to produce a valuable motor fuel component from low cost raw materials. The first decisive results were achieved by Hüls in the FRG, and Anic/Snamprogetti in Italy [3.74].

According to the Hüls MTBE process, the synthesis is carried out in the liquid phase at temperatures below 100 °C on an acid ion exchanger namely, a sulfonated copolymer made from styrene and divinyl benzene [3.75]. All C_4-hydrocarbons, including the C_4-acetylenes but with exception of the i-butene, behave very inertly under the reaction conditions. Hence C_4-cuts from steam crackers, before or after butadiene extraction, and also FCC-streams can be employed immediately. Per ton MTBE, 0.647 t i-butene and 0.392 t methanol are required. The MTBE contains 1.9% by wt. i-butene and 3.8% by wt. methanol. The methanol can be separated by an additional separation stage and recycled to the reaction. The specific methanol consumption then falls to 0.366 t [3.76]. A flow chart of the process is given in Fig. 3.19.

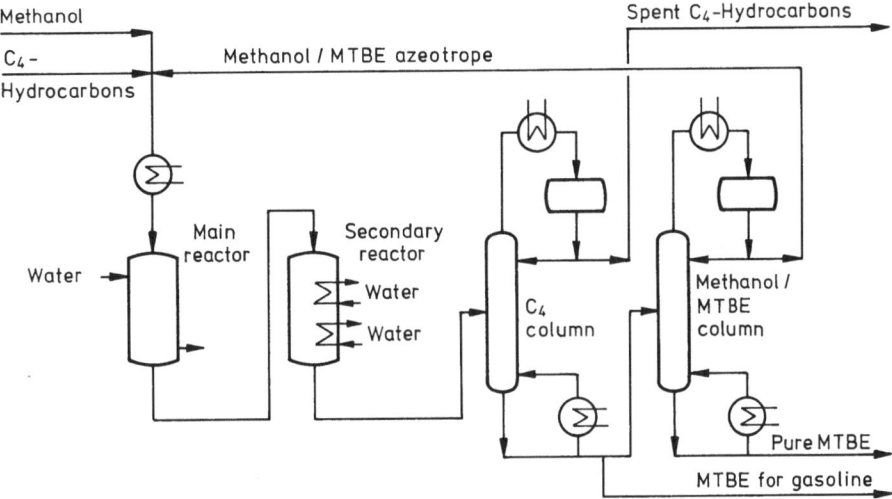

Fig. 3.19. Hüls MTBE process [3.76]

Fig. 3.20. Hüls MTBE plant

Table 3.6. World MTBE capacities in 1986 and additional capacities up to the end of 1987 from plants under construction (in 1000 t/a) [3.80]

Area, Country	Company	Site	Process	Capacity 1986 up to 1987
				3345 + 705
N. America				
USA	Amoco	Whiting, IN	Chem. Res. and Lic.	125
		Yorktown, VA	Chem. Res. and Lic.	20
	Arco	Channelview, TX	Arco	750*
		Channelview, TX	Arco	400
		Corpus Christi, TX	Arco	500*
	Champlin Petroleum	Corpus Christi, TX	Hüls	85
	Diamond Shamrock	Sunray, TX	Arco	95
	Exxon	Baytown, TX		125
	Final Oil and Chemical[1]	Big Spring, TX		23
	Hill Petroleum[2]	Houston, TX	Chem. Res. and Lic.	70
	Phillips	Sweny, TX	Phillips	143
	Sun Refining and Marketing	Marcus City, PA	Arco	315 + 155
	Texaco	Port Neches, TX	Texaco	280 + (80 TAME)
	Texas Petrochemicals	Houston, TX	Snamprogetti	550** + 550**
		Houston, TX		64
	Valero Refining	Corpus Christi, TX		(150)
	Transamerica Refining	Good Hope, LA		130 + 40
S. America				
Argentine	Petrochimica Gen. Mosconi	Ensenada		+ 40
Brasil	Comp. Petrochimica do Sul	Triunfo Copene	Petroflex	70
	Copene			60
W. Europe				
Austria	Petrochemie Danubia	Schwechat	Snamprogetti	679 + 243
Belgium	SIBP	Antwerp	Phillips	50
Finland	Neste Oy	Porvoo	Snamprogetti	80
France	Elf	Feyzin	IFP	80
FRG	Hüls	Marl	Hüls	150 + 40
	Deutsche Texaco	Heide	Texaco	15

Country	Company	Location	Process		
Greece	Hellenic Aspropyrgos	Krioneri	Snamprogetti	+	63
	Motoroll Hellas	Corinth	Snamprogetti		34
Italy	Enichem	Ravenna	Snamprogetti		125
	Selm	Priolo	Hüls	+	40
Netherlands	DSM	Heerlen	Snamprogetti		75
	Shell	Pernis	Shell		125
Spain	Petronor	Somorrostro	Hüls		45
United Kingdom	Lindsey Oil[4]	Immingham	Phillips	+	100
				(+	50 TAME)
E. Europe					
CSSR	Chempetrol	Kralupy	Hüls		171
GDR	Leuna Werke	Leuna	Hüls		91
Yugoslavia	FSK	Zrenjanic	Snamprogetti		45
					35
Remaining World					
China	Qilu Petrochemical				110
Israel	Dor Petrochemicals		IFP	+	90
				+	40
					30
Japan	Mitsui	Chiba	Hüls		80
Singapore	Petrochem Corp. of Sing.	Pulau	Sumitomo	+	50
				4735 +	1078

* via butane oxidation
** via butane dehydrogenation

[1] formerly Cosden Oil
[2] formerly Charter Oil
[3] closed down
[4] joint venture Total/Petrofina

According to the equilibrium constant of the reaction of i-butene with methanol, a stoichiometric charge attains only 92% conversion. To achieve a higher i-butene conversion requires an excess of methanol, which can usually be distilled from MTBE as an azeotrope with low methanol content (14% by volume). In the Hüls process rectification is performed under pressure (8 bar) which reduces the MTBE content of the circulating azeotrope by about a half. i-Butene conversions of 98% and MTBE-purities of >99.7% (fine chemical quality) can be realised using a secondary reactor and recycle procedure. Hüls began commercial production with the first plant, which had a capacity of 60,000 t/a MTBE, in 1976. In 1978, this was replaced by a new plant with a capacity of 150,000 t/a MTBE (commercial name — Driveron). A photograph of this plant, which is still in operation, is shown in Fig. 3.20. Another view can be seen on the book cover.

Hüls has already awarded several licenses all over the world for their process.

The first MTBE plant in the world went into production in 1973 at Anic in Ravenna, using a similar process from Snamprogetti/Anic [3.77]. Most of the world MTBE licenses have been awarded for this process. The Deutsche Texaco feeds, in Heide, the smallest MTBE plant in the world (15,000 t/a) with butadiene-containing C_4-cuts, based on its own process. This proved to be very economical there because the C_4-cut was thus reduced by the removal of the i-butene contents, so the transport of the residual C_4-cut to EC Cologne-Worringen, for butadiene extraction proved to be less expensive.

As a result of surplus capacity in the refineries part of their distillation plants, they can, in many cases, be used for producing MTBE without large constructional changes. Chemical Research and Licensing Comp./Neochem. Corp. and IFP license MTBE processes especially on the basis of refinery B-B [3.78].

The present producers of MTBE and their capacities are listed in Table 3.6.

The world's largest producer of MTBE is Arco, with a total capacity of 1.65 million t/a, entirely in the USA. Arco obtains the major part from i-butane (Channelview, 750,000 t/a; Corpus Christi, 500,000 t/a). The i-butane is first converted to TBA (Sect. 2.3.2). Its dehydration to i-butene and the addition of methanol to yield MTBE on suitable catalysts in a reactor are claimed to have processing costs similar to those of the preparation of MTBE from raffinate I [3.79]. From the beginning of 1988, Arco in the Netherlands will produce 370,000 t/a MTBE, and probably from the end of 1988 in France, 420,000 t/a MTBE, both using this process [3.80].

The Etherol process of BP may be considered to be related to MTBE production. In this process, however, not only the single raw material i-butene alone or within a C_4-cut but the reactive olefines in C_4-C_6-streams are converted together with methanol to high octane gasoline components. The products vary with the process version and feed. They can be MTBE, *tert*-amyl mehtyl ether (TAME), TAME gasoline (i. e., hydrocarbon stock containing TAME), mixed ethers (e. g., FCC gasoline enriched with MTBE, TAME and higher ethers), or various combinations [3.82].

Figure 3.21 gives a flow chart of the Etherol process using a FCC naphtha cut. The feed passes through a guard reactor to remove impurities and is then converted with methanol over a special acid resin catalyst in two fixed-bed reactors at low temperatures and pressures. The gasoline produced with mixed ethers is withdrawn over a heat exchanger. Afterwards a special plant section is provided to recycle unreacted

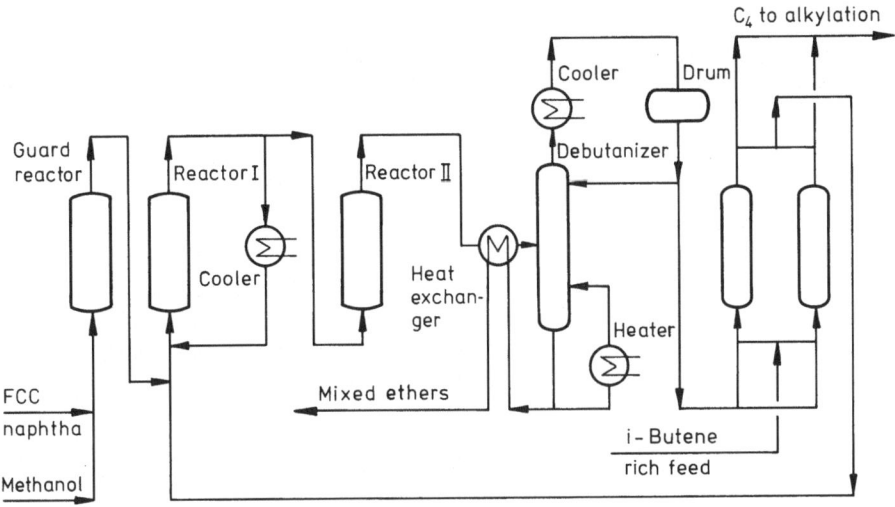

Fig. 3.21. BP Etherol process [3.82]

methanol and to separate a C_4-cut for alkylation. A fresh C_4-feed is used in the separation section as a purging medium for methanol. Alternatively a water wash for methanol removal can be provided [3.82].

The first commercial plant with a capacity of 360,000 t/a mixed ethers was commissioned early in 1986 at the BP Vohburg refinery.

3.3.2 Chemical Secondary Products from Raffinate I

3.3.2.1 Polybutene

Polybutene can be obtained from raffinate I using the low-temperature Friedel-Crafts catalysis (processes of Amoco and Cosden Oil). Refinery butenes are preferred in the USA as the raw material; steam cracker butenes are the principal starting material in Europe.

The crude C_4-stream is first pretreated mainly in order to remove sulfur components and water. Then follows a liquid phase polymerisation in the reactor at temperatures between -10 and $-80\,°C$, in the presence of an aluminum chloride catalyst; there reacts principally i-butene, plus a few percent of n-butenes and i-butane.

The molecular weight of the polybutene varies between 300 and 3000, depending on the cooling conditions and catalyst concentration. It is a disadvantage that only 80 to 95% of the i-butene polymerises and that therefore the purity of raffinate II is unsatisfactory. The spent catalyst is separated from the product in a settling tank and residual amounts are removed by washing with alkali and water. The polybutene is finally purified by distillation.

The plant equipment can be used with such flexibility that poly-i-butene and butyl rubber can be prepared batchwise in alternation. The wide range of the degree of polymerisation can yield polybutene products for a broad spectrum of applications. These range from lube oil additives, film sealants, sealing compounds, and adhesives to plastic products such as film materials, protective coatings and cable insulating materials.

In the FRG only poly-i-butene of higher qualitative grades has so far been produced (see Sect. 3.3.4.1).

3.3.2.2 Di- and Tri-i-butenes

It was formerly customary to remove i-butene from the cracking cuts by oligomerisation, in order to obtain n-butene for butadiene production by dehydrogenation. This type of separation was frequently practised when there was no outlet for monomeric i-butene. The hydrogenated oligomerisate has high octane numbers (RON 100 to 115) and was thus regarded favourably as a useful component for the gasoline pool. Recently, however, chemical intermediates are being based on these butene oligomers.

Oligomerisation of the i-butene is performed according to the classical Cold Acid process using 60 to 65% sulfuric acid at 20 to 25 °C. Under these mild conditions only the i-butene contents of a butene mixture react to yield the oligomers di- and tri-i-butene (DIB, TIB). When the conditions are more severe (60 to 75 °C) codimers or cotetramers are formed together with the n-isomers [3.83]. Before their MTBE plant began production in 1978 Hüls were obliged to carry out i-butene oligomerisation according to their own Cold Acid process (28,000 t/a) in order to obtain pure 1-butene, among other products, for production of poly-1-butene when necessary.

In the Bayer process the butene mixture is oligomerised in a 1% suspension of a cationic ion exchanger at 100 °C [3.84; 3.85], thereby converting 99% of the i-butene to the oligomers DIB and TIB. About 10% of the n-butenes react during normal operation to codimers or cotetramers. Owing to favourable differences in boiling point, distillation yields C$_8$- and C$_{12}$-olefines of >98% purity and C$_{16}$-olefines of >95% purity. It is even technically possible to separate codimers. EC-Cologne-Worringen has been operating a 100,000 t/a plant according to this process since 1961. From this amount 32,000 t/a can be hydrogenated to i-octane/i-dodecane.

DIB is also yielded as a by-product, namely of the Dimersol X process of IFP. Since this process primarily serves for deriving straight-chain plasticiser oligomers, the preferred raw material is raffinate II which is rich in n-butenes. Only very small amounts of DIB are then formed. On the other hand the procedure is so flexible that even raffinate I can be used as a feed. Nissan in Kashima/Japan, using a raffinate I containing 47% i-butene by wt. obtained 6.8% of DIB [3.86].

The separation of DIB by distillation is obligatory if it is desired to prepare plasticiser alcohols by the OXO process from the n-C$_8$- and n-C$_{12}$-olefines. This additional purification requirement impairs the economic success. Furthermore, it must be borne in mind that IFP collects licensing fees which are based on the proceeds for the plasticiser olefines; these are usually evaluated higher than the gasoline components.

3.3.2.3 Propene

Phillips and IFP are developing processes for the metathesis of ethylene/2-butene mixtures into propene [3.29]. The Phillips process is performed at high temperatures, using a molybdenum oxide catalyst. The IFP process, in contrast, at lower temperature used a different catalyst. Lyondell, a subsidiary company of Arco, began production in Channelview, TX in 1985 with a plant using the Phillips process, for 135,000 t/a propene. The 2-butene is obtained there by dimerisation of ethylene [3.87]. A pilot plant of the IFP process went into production in Taiwan at the end of 1987.

To meet the universally increasing demand for pure propene as a result of the polypropene boom these processes might be helpful. They compete with production of propene in refineries, based on the, as yet, unexploited potential of the catalytic cracking plants (propene splitting).

3.3.2.4 Specialities

The specialities from raffinate I are essentially derivatives of di- and tri-i-butenes. Thus DIB is converted in the OXO-synthesis to the respective plasticiser alcohol, i-nonanol. Alkylation of phenol with DIB gives the detergent intermediate p-tert-octylphenol, and with TIB, dodecylphenol, which can also be prepared from tetra-propene and phenol. It is a reaction component for special phenolic resins. Bayer prepare tert-dodecylmercaptan (TDM) from a special TIB-cut (B. P. between 175 and 182 °C); it is not a single substance but a mixture of various isomeric thiols of the general composition $C_{12}H_{25}SH$. For this, the TIB-cut is mixed with hydrogen sulfide in the mole ratio of 1:3 and the reaction carried out in the presence of boron trifluoride [3.88]. TDM is a constituent of special styrene-butadiene- and nitrile-rubbers and the corresponding latices. Bayer has a TDM-capacity of 1,800 t/a.

3.3.3 Chemical Secondary Products from Raffinate II

3.3.3.1 Oligomers of n-Butenes

Straight-chain C_8-C_{12}-olefines and -alcohols have recently become available from raffinate II. They were formerly produced only from ethylene by Ziegler-synthesis sometimes combined with the OXO-synthesis. One can also make use of those heptenes which are more branched with methyl groups, yielded during the production of polymer gasoline (see Sect. 3.3.1.1).

Linear dimers can be obtained selectively from propene- and n-butene-containing cuts by means of the Dimersol X process of IFP [3.86, 3.89]. The raw materials must be free of diolefines and acetylenes. This is accomplished by selective hydrogenation. Drying is also necessary. A soluble Ziegler catalyst system is used in a series of reactors for this dimerisation in the liquid phase. At temperatures of 40 to 50 °C and only slight pressure, no corrosion has been observed. The spent catalyst is separated, neutralised and discarded. The non-reacted hydrocarbons in the product stream are not recycled to the reaction but are separated by distillation, the oligomers then being removed from the bottom. Yields of about 65% by wt. of n-octenes (DNB) and methylheptenes are obtained from raffinate II with 75% by wt. n-butene content [3.90]. This process is a modification of that developed for refineries to be used there for propene dimerisation to i-hexenes with high octane numbers (Dimersol G).

The first Dimersol X plant for 20,000 t/a DNB went on stream at Nissan in Kashima, Japan at the end of 1980. It supplies an OXO plant in Chiba where plasticiser alcohols are made.

BASF has also been producing DNB in Ludwigshafen since 1985, using the Dimersol X process from raffinate II. There is said to be a considerable capacity.

The Octol process has been developed recently with the German-American cooperation of Hüls and UOP [3.91]. A heterogeneous, long-lasting catalyst is used, in contrast to that of the Dimersol process, and the unreacted hydrocarbons can be recycled, thereby increasing the yield of n-octenes (DNB) to about 80% [3.92]. It is also possible to vary process conditions so that more n-trimers (TNB) are formed. These have good prospects as starting materials for preparing α-olefinesulfonates or C_{13}-oxoalcohols in the detergent sector.

The first commercial unit of the Octol process was completed in Marl in 1983 and has a capacity of 10,000 t/a. It employed a catalyst which yielded more branched products according to the market situation at that time (Octol A). This catalyst was later replaced by a new type on which oligomers with the desired high degree of linearity were produced (Octol B). This improvement is shown in Table 3.7 by comparing the iso-index of the olefines from the different sources. The iso-index is a measure of the average number of methyl groups in the side-chains of the molecule. Thus n-octenes have an iso-index of zero and dimethylhexene has an iso-index of two. A mixture of C_8-olefines containing 50% n-octenes and 50% methylheptenes would have an iso-index of 0.5 [3.22].

In the autumn of 1986 a licensed Octol plant commenced production (45,000 t/a) in Sakai, Japan at the General Sekiyu Refineries (GSK), a subsidiary company of Exxon, based on FCC C_4-hydrocarbons [3.79]. A flow chart of the Octol process is given in Fig. 3.22.

Table 3.7. The iso-index of C_8-olefines from diffe-rent sources [3.22]

Source of C_8-olefines	iso-Index
Shell SHOP process	0.0
Hüls Octol B process	1.1
IFP Dimersol X process	1.3
Hüls Octol A process	1.9
UOP Polymer gasoline dimer fraction	2.1

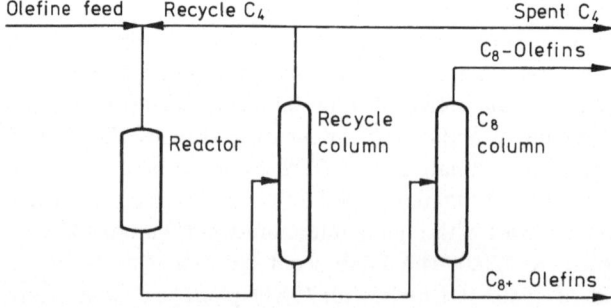

Fig. 3.22. Hüls/UOP Octol process [3.22]

3.3.3.2 Butadiene

Butadiene can be obtained from raffinate II or *i*-butene-free refinery B-B by dehydrogenation or oxydehydrogenation. The older processes require the removal, as far as possible, of the butanes from the *n*-butenes.

In the temporarily dominating process of Standard Oil the *n*-butenes were preheated and dehydrogenated in shaft reactors using superheated steam as heat carrier. Unreacted butenes were recycled after the water has been condensed and the butadiene extracted [3.93]. The most important catalysts and reaction conditions are quoted in Table 3.8.

The Dow catalyst was the most favoured in practice although it had to be regenerated frequently.

In the USA in 1967 butene dehydrogenation plants existed at: Copolymer (70,000 t/a) in Baton Rouge, LA; Enjay (Standard Oil, 30,000 t/a) in Baytown, TX; Neches Butane (joint venture of Gulf Oil, B. F. Goodrich, Texaco, and U. S. Rubber, 325,000 t/a) in Port Neches, TX; and Petroleum Chemicals (Cities Service, 70,000 t/a) in Lake Charles, LA [3.94]. In Western Europe there was one butene dehydrogenation plant, that of British Esso in Fawley, only in operation for a short time.

Adding atmospheric oxygen (oxydehydrogenation) has a doubly positive effect on butene dehydrogenation. The exothermic reaction of the oxygen with the split off hydrogen displaces the dehydrogenation equilibrium in the direction of the desired products, thereby markedly increasing the conversion or throughput of the plant. Moreover the thermal balance is improved because of lower steam consumption.

The most important oxydehydrogenations are the Oxo-D process of Petro-Tex and the O-X-D process of Phillips.

Using a long-life catalyst which does not need regeneration, conversions of 70% with butadiene selectivity of 96% can be obtained in the Oxo-D process. The concentration of butadiene in the product stream is about 58% [3.96]. Petro-Tex built a large scale plant (270,000 t/a butadiene) in Houston, TX, USA in 1965 using this process. This plant is still in operation under the changed name of Texas Petrochemicals.

Conversions of about 80% with selectivities of 90% are attained in the O-X-D process. The butadiene concentration is said to be 63% [3.97]. Phillips have been running a large scale plant for 140,000 t/a butadiene in Borger, TX since 1971.

Table 3.8. Catalysts and reaction conditions of butene dehydrogenation [3.95]

Reaction conditions		Catalysts		
		Phillips No. 1490, Fe_2O_3 on bauxite	Shell No. 205, Fe_2O_3	Dow type B Ca-Ni-phosphate + Cr_2O_3
Temperature	°C	620 to 680	620 to 680	600 to 680
Pressure	bar	1.5 to 1.8	1.5 to 1.8	1.2 to 2
Mole ratio, H_2O/butene		9 to 12/1	8/1	20/1
Space velocity	vol/vol. h	300 to 400	approx 500	125 to 175
Conversion per pass	%	27 to 33	26 to 28	<45
Selectivity for butadiene	%	69 to 76	73 to 75	90
Regeneration (duration, frequency)		—	1 h/24 h	15 min/30 min

BP Chemicals and Dow have also developed other processes for oxydehydrogenation.

3.3.3.3 Methyl Ethyl Ketone

About 90% of the world production of methyl ethyl ketone (MEK) comes from *sec*-butanol (SBA), which is available from the *n*-butenes of raffinate II via indirect or direct hydration (see Sects. 3.1.2.2 and 3.1.2.4).

Cu-, Zn- or bronze-catalysts are suitable for the dehydrogenation in the gas phase of SBA to MEK:

$$CH_3-CH_2-\underset{\underset{OH}{|}}{CH}-CH_3 \xrightarrow{-H_2} CH_3-CH_2-\underset{\underset{O}{||}}{C}-CH_3$$

At present, the sole producer in the FRG is the Deutsche Texaco plant in Moers-Meerbeck, with a capacity of 60,000 t/a SBA, used almost completely for producing MEK. Conversions of 90 to 95% are attained there in the dehydrogenation at 240 to 260 °C, at normal pressure, and using a precipitated copper-containing catalyst. Less than 5% butenes and higher ketones are formed. The catalyst must be replaced after 3 to 4 months and reactivated by oxidation [3.98]. The plant is flexible and can be used to prepare acetone by dehydrogenation of *i*-propanol in an analogous way.

The remaining MEK, about 10% of the world production, was obtained until recently as a co-product during the manufacture of acetic acid by catalytic oxidation of *n*-butane (see Sect. 3.2.3).

The direct oxidation of *n*-butenes according to the Hoechst-Wacker process, using $PdCl_2/2$ CuCl as redox catalysts, follows the simplified equations [3.99]:

$$CH_3-CH_2-CH=CH_2 + PdCl_2 + H_2O \longrightarrow CH_3-CH_2-\underset{\underset{O}{||}}{C}-CH_3 + Pd + 2HCl$$

$$Pd + 2CuCl_2 \rightleftharpoons PdCl_2 + 2CuCl$$
$$2HCl + 1/2 O_2 \longrightarrow H_2O + Cl_2$$
$$2CuCl + Cl_2 \longrightarrow 2CuCl_2$$

The yield of MEK is said to be about 85%. A disadvantage is the formation of co-products (10%) consisting of an impure acidic organic mixture (*n*-butanal, chlorinated products, hydrochloric acid). Problems arise caused by both their corrosive properties and the difficulty of disposal with respect to the environment. Evidently these are the reasons why the process has not been put into operation on a large scale.

The MEK production in the western world was estimated to be 700,000 t in 1983 [3.100].

3.3.3.4 Maleic Anhydride

Maleic anhydride (MA) can also be produced by oxidation of raffinate II, analogous to the oxidation of *n*-butane. Any butadiene present is also converted. Modified V/P catalysts are employed:

$$CH_3-CH_2-CH=CH_2 + 3 O_2 \longrightarrow \begin{matrix} CH-C \\ || \quad \diagdown O \\ CH-C \end{matrix} + 3 H_2O$$
$$CH_2=CH-CH=CH_2 + 5/2 O_2 \longrightarrow + 2 H_2O$$

The first oxidation plant yielding MA from butene was put into operation in 1962 by Petro-Tex in Houston, TX [3.101]. It was sold to Denka at the beginning of the 1970s and converted to benzene oxidation with a capacity increase to 23,000 t/a MA. The plant was reconverted to butane oxidation (20,000 t/a MA) at the beginning of the 1980s. This demonstrates the way in which the butene procedure can easily be adapted to the raw material situation.

When butenes are oxidised the reaction heat is relatively large (18,000 kJ/kg). For the same throughput the reactor must thus have about twice the heat exchange surface area and the cooling equipment produce twice the performance needed when using benzene or butane as raw material.

There is no doubt that, when it is available, n-butane is the cheaper raw material. Nevertheless at the beginning of the 1970s some MA plants based on raffinate oxidation were built, especially in the FRG and Japan. A semicommercial plant (3600 t/a) with tube bundle reactor and solid bed catalyst was operated satisfactorily at BASF in the period 1969−1974 [3.102]. Licences for the BASF MA process were granted for the construction of two plants in Japan, by Dai Nippon (18,000 t/a) and Nichiyu (6000 t/a). The latter is still working today. Bayer also has a capacity of 10,000 t/a MA using a similar process which has been operated since 1969 in Krefeld-Uerdingen [3.103]. Raffinate II with 75% n-butene content by wt. is used yielding 62% by wt. of pure MA, referred to raffinate II. A last example is Mitsubishi which has been producing MA in Mizushima since 1970 in a plant equipped with a newly developed fluid bed tube reactor; the MA capacity there is 18,000 t/a [3.104].

In addition to the Bayer plant mentioned above, which is based on butene, there are two further large scale MA plants based on benzene in the FRG: Deutsche Texaco, Moers-Meerbeck (8000 t/a) and Hüls, Bottrop (30,000 t/a). Additional MA is obtained by reprocessing the wash water for waste gas from the phthalic anhydride production by oxidising o-xylene. In the FRG the quantities come from BASF, Ludwigshafen (up to 3000 t/a), Bayer, Leverkusen (up to 2000 t/a), and Hüls, Bottrop (up to 3000 t/a).

3.3.3.5 Acetic Acid

The introduction of the oxidation of butane and naphtha for preparing acetic acid in the 1960s induced several firms in Europe and Japan to search for a suitable oxidation process which would make use of raffinate II [3.46, 3.105]. Hüls and Bayer in the FRG developed an oxidation process with atmospheric oxygen which operated at temperatures around 270 °C using a Ti/V-catalyst. The yield of acetic acid was 70% at a conversion rate of 73%. Unavoidably, 25% carbon oxides were also yielded [3.106].

The high steam supply needed to reduce the danger of explosion was a disadvantage. The acid yielded was diluted thereby so that additional energy was necessary for concentration.

Despite these deficiences, Bayer, Dormagen decided to build a plant for producing 11,000 t/a acetic acid. It was in operation, however, for only a short time. Parallel to this Bayer developed a two-stage process. Raffinate II was first converted into sec-butyl acetate on an acid ion exchanger at 100 °C and 20 bar, using circulating acetic acid. The ester was then oxidised to acetic acid at 200 °C and 60 bar, without a catalyst. Yields of 58% of acetic acid were obtained but here, too, 28% carbon oxides were yielded [3.106]. No attempt has yet been made to commercialise the process. Developments in other countries have not encouraged acetic acid producers to employ *n*-butene oxidation.

Attempts to oxidise propene, excess of which was previously available as a co-product, to acetic acid were also unsuccessful. Patents for this propene oxidation referred in fact to "yields" of 50% CO_2! There is no doubt about the present and future superiority of the acetic acid processes based on synthesis gas, i. e. the carbonylation of methanol using the high pressure process of BASF and especially the low pressure process of Monsanto. Numerous plants also exist for the one- or two-stage Wacker process, an oxidation of ethylene via acetaldehyde. BP is applying direct oxidation processing of naphtha to acetic acid, where formic and propionic acids are yielded as co-products. Reference is made to Sect. 3.2.3 for the preparation of acetic acid by oxidation of *n*-butane.

3.3.3.6 Specialities

At present no *n*-butene speciality requires pure *n*-butenes. Raffinate II is sufficient as a raw material.

2,3-Butanediol is an important speciality for making polyurethanes (e. g. Vulkollan of Bayer from the reaction with naphthalene-1,5-diisocyanate) and an intermediate of technical importance in the pharmaceutical and plant-protection fields (the insecticide Sapecron from Bayer). To prepare 2,3-butanediol, raffinate II is treated with chlorine water according to the chlorohydrin process, and the product converted with NaOH to a mixture of about 85% 2,3-butene oxide and 15% 1,2-butene oxide. Hydrolysis of this mixture at 160 to 220 °C under 50 bar pressure yields the corresponding diols which can more easily be separated by distillation than can the epoxides [3.107].

The OXO-synthesis converts raffinate II butenes into a mixture of about 70% pentanal (*n*-valeraldehyde) and 30% 2-methyl-1-butanal. These can be hydrogenated to the solvents 1-pentanol (*n*-amyl alcohol) and 2-methyl-1-butanol, respectively, which can again be separated by distillation.

It would also be possible to dehydrate methylbutanol to methylbutenes which are easily dehydrogenated to isoprene. In Europe, Ruhrchemie, Oberhausen, produces *n*-amyl alcohol, among other compounds in its OXO-plant. The sole producer in the USA is Union Carbide in Seadrift, TX, since Dow recently closed down its production plant for amyl alcohol in Freeport, TX.

The alkylation products of phenol using oligomers of *n*-butene are also specialities. They are made in the FRG by Hüls.

3.3.4 Chemical Secondary Products from Pure *i*-Butene

Pure *i*-butene is here understood to be high-purity material only available at present from decomposing *tert*-butanol, MTBE, or *tert*-butylphenols. Specifications for the required purity are given in Table 1.5 in Sect. 1.4.

3.3.4.1 Poly-*i*-butene

Pure *i*-butene is polymerised in an inert solvent at temperatures between −10 and −100 °C, yielding the rubber-like product, poly-*i*-butene, which alone cannot be vulcanised [3.108]. The process and catalyst correspond to the previously described production of polybutene (see Sect. 3.3.2.1). Esso Chemie produces, in Cologne, under the trade name of Vistanex, up to 35,000 t/a poly-*i*-butene (PIB) of intermediate molecular weight for use as a lubricant additive especially for improving the viscosity of lubricating oils. The same plant was also used to produce DIB/TIB-cuts as gasoline components up to 1980, when this procedure on the basis of pure *i*-butene was no longer competitive. BASF has a PIB capacity in Ludwigshafen of 10,000 t/a and at ROW, Wesseling, of 8000 t/a. The product types have a wide range of molecular weight and are sold under the name of Oppanol B. Rhepanol waterproof sheathing is a secondary product of PIB as are filling materials. A product sample of Oppanol B is shown in Fig. 3.23.

3.3.4.2 Butyl Rubber

Copolymerisation of pure *i*-butene with 1 to 3% isoprene yields butyl rubber (IIR). The technical polymerisation process is similar to that for making poly-*i*-butene. The product can be vulcanised to some extent and is characterised by its impermeability to gases amongst other properties. Tubeless tyres are therefore coated internally, mainly with IIR. No IIR is produced in the FRG. Table 3.9 summarises the information about present producers and their capacities in Western Europe. Esso produces halogenated IIR in batches as a speciality. This material is more easily processed and less susceptible to aging.

Fig. 3.23. Product sample of Oppanol B

Table 3.9. Production of butyl rubber in Western Europe, 1984 [3.109]

Country	Company, site	IIR capacity in 1000 t/a
Belgium	Polysar, Antwerp	85
France	SOCABU*, Port Jerome	46
United Kingdom	Esso, Fawley	50
Total		181

* joint venture of Esso Chemie (80%) and ATO (20%)

The Italian company Pressindustria has recently established a joint venture investment at Petrochemiecomplex Tobolsk in Siberia, USSR. A production plant for 90,000 t/a butyl rubber is planned, of which 60,000 t/a are to be the regular product and 30,000 t/a chlorinated butyl rubber. Two thirds of the production is to be used in the Soviet Union automobile industry and one third for export. It is scheduled to go into operation in 1991 [3.110].

3.3.4.3 Methacryl Compounds

Methyl methacrylate (MMA) is the monomer for the plastic product polymethyl methacrylate (PMMA). World capacities for PMMA are shown in Table 3.10. Predominant in the world production of MMA has so far been the acetone cyanohydrin process, based on development work of Röhm in Germany and ICI in Great Britain. There are three stages:

Acetone and hydrocyanic (prussic) acid yield the cyanohydrin which is converted to methacrylamide with concentrated sulfuric acid. Hydrolysis or methanolysis gives methacrylic acid or MMA, respectively [3.112]. After the last reaction step the product separates into an organic upper phase containing the MMA and an aqueous, lower phase containing the sulfuric acid and ammonium hydrogen sulfate. After MMA has been separated, the lower phase is generally neutralized with ammonia to give ammonium sulfate.

This classical route is applied by companies who have hydrocyanic acid at their disposal. The two producers of MMA in the FRG prepare their own hydrocyanic acid; Röhm in Darmstadt uses the Andrussow process:

$$CH_4 + NH_3 + 3/2\ O_2 \longrightarrow HCN + 3H_2O$$

and Degussa in Wesseling the BMA process:

$$CH_4 + NH_3 \longrightarrow HCN + 3\ H_2$$

In addition hydrocyanic acid is a co-product of the production of acrylonitrile from propene by ammonoxidation according to the Sohio process of Standard Oil of Ohio, USA. This is carried out in the FRG by Bayer at EC, Cologne-Worringen (200,000 t/a) and Hoechst, Münchsmünster (110,000 t/a). The use of a uranium oxide catalyst yields about 8% by wt. of co-product hydrocyanic acid; with older Mo-Bi-catalysts, up to 25% has been formed, referred to the acrylonitrile [3.113].

Table 3.10. World production of methyl methacrylate, 1988 [3.111]

Country, company, site		MMA-capacity in 1000 t/a	Raw material (except methanol)
USA			
Rohm and Haas, Deer Park, TX		340	
DuPont, Memphis, TN		110	acetone,
Belle, VA[1]		(55)	HCN
Cyro Industries[2], Fortier, LA		90	
		540	
W. Europe			
France	Norsolor, Saint Avold	60	acetone, HCN
FRG	BASF, Ludwigshafen	36	ethylene, CO, CH_2O
	Röhm, Worms	115	
	Degussa, Wesseling	45	
Italy	Vedril[4], Rho	50	acetone,
Spain	Paular, Tarragona	30	HCN
	Monacril, Palos de la Frontera	20	
United Kingdom	ICI, Billingham	105	
	Rohm and Haas, Teesside[3]	(35)	
		461	
Japan			
Mitsubishi Rayon, Ohtake		75	acetone,
Asahi, Kawasaki		?	HCN
Mitsubishi Rayon, Ohtake		40 + 55	TBA
Nihon Methacryl Monomer[5], Ehime		40	i-Butene
Asahi, Matsushima/Kawasaki		?	TBA
		~ 300	
Remaining world (Brazil, Mexico, India, Taiwan)		~ 70	
World total		ca. 1370	

[1] plant produces no MMA but methacrylic acid and higher methacrylates
[2] joint venture of American Cyanamide and Röhm (FRG)
[3] plant closed down since the beginning of the 1980s
[4] subsidiary company of Montedison
[5] joint venture of Nippon Shokubai and Sumitomo

A high yield (>90%) of highly pure MMA is furnished by the ACH process. The technology has matured since the problems of corrosion were overcome. The neutralisation of the unavoidable amounts of ammonium hydrogen sulfate is a disadvantage, however. The output of ammonium sulfate amounts to about 2.2 t per t of MMA and is, moreover, contaminated with organic substances. Since this impure ammonium sulfate is a troublesome residual material, the two firms in the FRG which produce MMA have abandoned the neutralization procedure. Instead, they thermally decompose the hydrogen sulfate with the excess waste sulfuric acid (together giving 2.5 to 3.3 times the amount of MMA) at about 1000 °C. This expensive process yields nitrogen, water and SO_2. The last two components are reprocessed to sulfuric acid.

Other efforts to circumvent this problem of the waste by new routes have so far had little success. Figure 3.24 depicts the syntheses of MMA. Oxidation processes of i-butene have been the focus of research interest during the last few years.

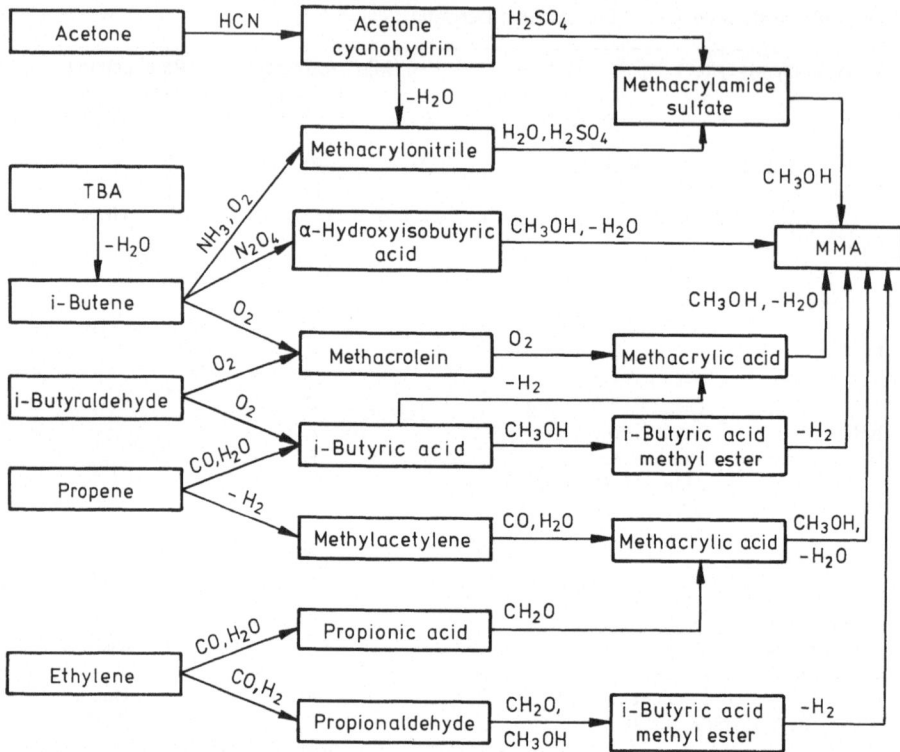

Fig. 3.24. Possible syntheses of methyl methacrylate [3.112]

The ammonoxidation to methacrylonitrile (Sohio process) and its alcoholysis, analogous to the ACH process mentioned above, have been achieved out technically. Asahi in Japan is running this process in a modified acrylonitrile plant in Mazushima, using TBA as raw material [3.115]. The methacrylonitrile is then shipped to the Asahi ACH plant in Kawasaki. It is converted there to methacrylamide sulfate which gives MMA on esterification. The economic success of the process is restricted by a number of problems and is evidently realized in this special case only because old, depreciated or re-equipped installations are used. Among the disadvantages are the many stages of the process, the high acid consumption during the ammonoxidation (the acid acts as acceptor for the ammonium radical), and the consumption of ammonia which must be removed as ammonium hydrogen sulfate, a low value by-product.

The direct oxidation of i-butene to α-hydroxy-i-butyric acid (HIBA) also creates technical difficulties; it can be dehydrated to methacrylic acid with 95% yield:

$$\begin{array}{ccccc} CH_3 \\ CH_3 \end{array}\!\!> C=CH_2 \xrightarrow{+2O_2} \begin{array}{c} CH_3 \\ CH_3 \end{array}\!\!> COH-COOH \xrightarrow{-H_2O} CH_2=C \!\!<\begin{array}{c} CH_3 \\ COOH \end{array}$$

In 1965 Escambia Corp. (belonging to Air Products) was running a 9000 t/a plant in Pensacola, FL, USA, using N$_2$O$_4$ in nitric acid as oxidation agent (Escambia process). Explosion misfortunes led to closure of the plant after only two years of operation. A similar process has been developed using N$_2$O$_4$ in acetic acid at temperatures of 250 °C (Lonza process).

The derivation of HIBA via *i*-butanediol (*i*-butane glycol) have also been developed to a technical scale. *i*-Butanediol can be obtained directly from pure *i*-butene through a Wacker liquid phase synthesis with a redox catalyst [3.106]:

$$CH_3{>}C=CH_2 \;+\; O_2 \;+\; 2H_2O \;\longrightarrow\; 2\; CH_3{>}COH-CH_2OH$$

The *i*-butene is oxidised by Tl^{3+} which is reduced thereby to Tl^+; this is reconverted to Tl^{3+} with hydrogen chloride/oxygen in a copper chloride solution:

$$
\begin{aligned}
2\,TlCl \;+\; 4\,CuCl_2 &\longrightarrow 2\,TlCl_3 \;+\; 4\,CuCl \\
4\,CuCl \;+\; 4\,HCl \;+\; O_2 &\longrightarrow 4\,CuCl_2 \;+\; 2\,H_2O \\
\hline
2\,TlCl \;+\; 4\,HCl \;+\; O_2 &\longrightarrow 2\,TlCl_3 \;+\; 2\,H_2O
\end{aligned}
$$

There are some technical and economic disadvantages because of the corrosive media in this process.

Alternatively, *i*-butanediol can be made from pure *i*-butene via *i*-butene oxide. This oxide can be derived through classical chlorohydrin formation, as used for propene and *n*-butene oxidations, or through catalytic direct oxidation. The latter is a liquid phase process in acetic acid, water and THF at 70 °C using thallium triacetate as catalyst and giving 82% yields of *i*-butene oxide. At rather higher temperatures the oxide can be immediately hydrolysed to the glycol [3.106].

According to investigations of Arco, *i*-butanediol can be oxidised to HIBA in high yield in the liquid phase at 70 to 80 °C on Pt-containing catalysts [3.106].

The total yields of MMA from *i*-butene via HIBA are about 60% in all cases. General acceptance of all these processes is hindered, however, by the corrosive properties of the acid.

The present research situation suggests that the most promising route to oxidation of *i*-butene is via methacrolein.

Mainly Japanese companies have been showing interest in developing processes for oxidising *i*-butene or TBA via methacrolein. These include Mitsubishi Rayon, Asahi, Nippon Zeon (formerly Japanese Geon), Nippon Kayaku and also Halcon/SD in the USA.

This research work has benefited from knowledge derived from the oxidation of propene. *i*-Butene is therefore often combined with propene in the coverage of patents. The oxidation of propene to acrolein by atmospheric oxygen has been established technically for a long time. The latest generation of propene catalysts give also acceptable yields of methacrolein from *i*-butene and can be adapted with only small modifications (Fig. 3.25).

In order to obtain yields of up to 80% methacrolein, however, the space velocity must be limited in order to prevent premature deactivation of the catalyst. A contact time of 2 to 3 s and only 6% concentration of *i*-butene are tolerable. The space-time yields are then, however, unsatisfactory. If the space-time yield is increased to more than 500 g methacrolein per litre catalyst and per hour, acceptable economically, the yield falls inevitably to approx. 70%.

The air oxidation of methacrolein to methacrylic acid still poses problems. Mitsubishi Rayon achieved a breakthrough by developing heteropolymolybdate catalysts which also give good results in the oxidation of *i*-butyric acid to methacrylic acid. It took almost 20 years from the beginning of laboratory work to the large scale technical version in the first commercial plant which began production of 40,000 t/a

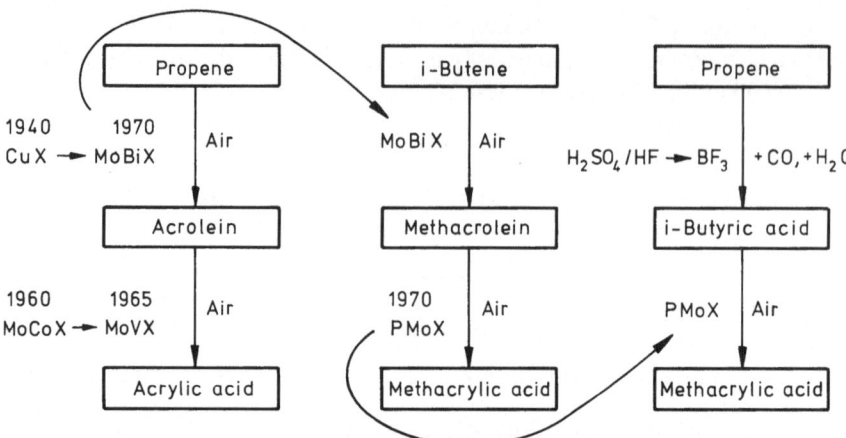

Fig. 3.25. Catalyst development for propene and butene oxidation processes

MMA in Japan in 1983. This plant of Mitsubishi Rayon is in Ohtake and obtains TBA as raw material from Mitsui Petrochemical Industries in Chiba [3.116].

TBA is directly oxidised with air in the vapour phase at temperatures of 300 to 400 °C, using a Mo-Bi-Sb catalyst. This yields a methacrolein mixture which is oxidised in the second stage to methacrylic acid, at 270 to 350 °C and in presence of a catalyst of the PMo-type. This is in turn esterified to MMA with methanol. The yield is said to be 68%, based on the stoichiometric *i*-butene content of the TBA [3.56]. Another MMA plant (40,000 t/a) is running in Japan according to a similar process of Nippon Shokubai at Nihon Methacryl Monomer in Niihama, a joint venture of Sumitomo Chemical and Nippon Shokubai [3.117]. This plant was built as a result of supply problems with hydrocyanic acid in Japan which made necessary a substitute for the Sumitomo ACH plant (formerly 17,000 t/a MMA). A flow chart of this process is given in Fig. 3.26 [3.118].

In the Nippon Shokubai process methacrylic acid (MAA) is manufactured in high yield from *i*-butene or TBA by vapour phase catalytic oxidation in two stages. In the first-stage reactor, methacrolein is formed which is further processed in the second-stage oxidation reactor to methacrylic acid without isolation. The product from this reactor is passed to a water scrubber to obtain an aqueous solution of MAA. Unreacted methacrolein in the product gas is recovered by an absorber and a stripper and can then be recycled to the second-stage reactor. MAA is removed from the aqueous solution with an organic solvent in an extractor. Part of the solvent is taken off by a separator. The MAA is continuously esterified in the liquid phase with methanol in a catalytic reactor. MAA is obtained in high purity by extraction and a two-stage distillation [3.117; 3.119].

In the USA, Arco in Bayport, TX and Rohm and Haas in Houston, TX announced the construction of MMA plants, each with a capacity of 137,000 t/a based on TBA or pure *i*-butene [3.120]. These plans have not yet been realized for economic reasons. Arco bought the world exclusive rights for the MMA process with TBA or *i*-butene from the Halcon/SD-group in summer 1987 and is thus holding the option for the commercialisation [3.121].

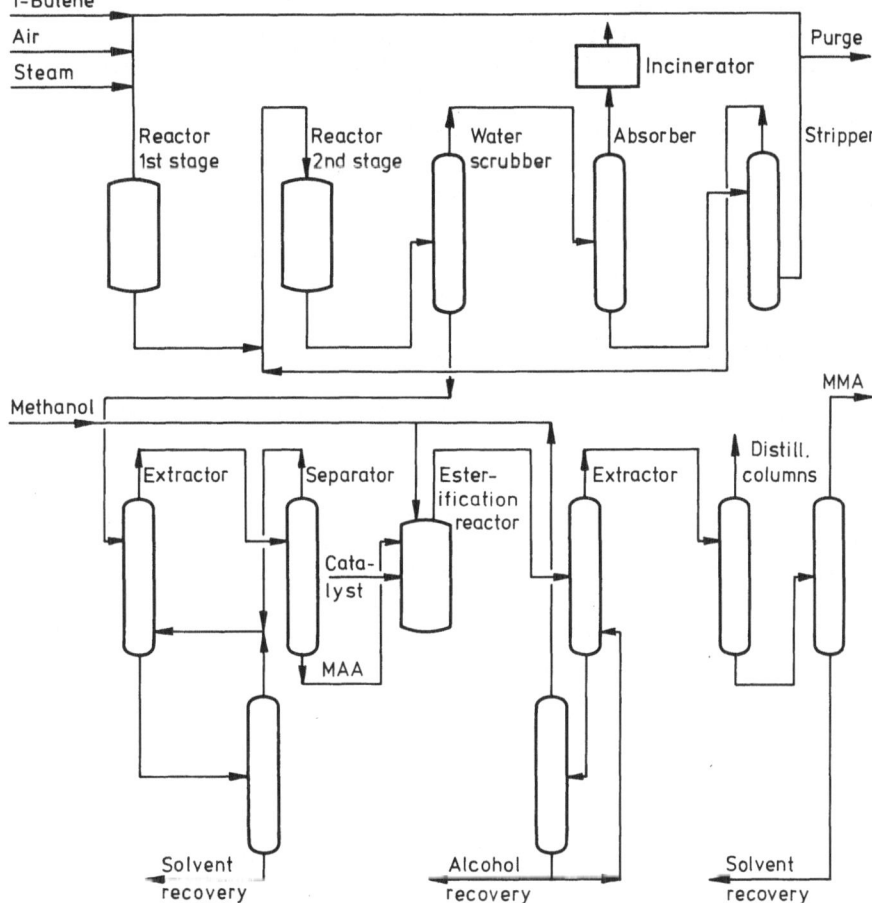

Fig. 3.26. Nippon Shokubai process for MMA production from *i*-butene or TBA [3.117; 3.119]

Technological development is proceeding in the direction of diminishing the reaction stages. The oxidative esterification of methacrolein can be mentioned here in this regard. The conversion is poor when the reaction conditions are outside the explosive limits. As a result there is an unacceptably high requirement for methanol recovery:

$$CH_2=C{<}^{CH_3}_{CHO} \;+\; CH_3OH \;+\; 1/2\,O_2 \;\longrightarrow\; CH_2=C{<}^{CH_3}_{COOCH_3} \;+\; H_2O$$

Another aim is still the single-stage oxidative esterification of *i*-butene (or butene mixtures or even LPG). Considerable R and D will have to be made to reach this goal:

$$CH_2=C{<}^{CH_3}_{CH_3} \;+\; CH_3OH \;+\; 3/2\,O_2 \;\longrightarrow\; CH_2=C{<}^{CH_3}_{COOCH_3} \;+\; 2H_2O$$

Processes for preparing methacrolein and methacrylonitrile from butadiene-free C_4-fractions without having to separate the butene are especially interesting. *i*-Butene and *n*-butene can be converted to methacrolein and butadiene, respectively, by oxidation of mixtures of them using suitable catalysts. The mixed ammonoxidation yields methacrylonitrile and butadiene in distinctly higher yields than when pure olefine components are used. This synergistic effect, not yet explained, has been reported in the literature [3.122].

Mitsubishi Rayon quotes in its patents high yields of up to 83.5% of methacrolein and 78.5% of butadiene; these data have not been verified up to now, however [3.59].

It is particularly difficult to obtain high conversion rates for the less energetically reacting *n*-butene and at the same time avoid complete oxidation of the *i*-butene to carbon oxides. It has thus been suggested that the mixed oxidation or ammonoxidation be interrupted at the point of maximum yield of the methacryl compounds and that the *n*-butenes and butadiene be oxidised to maleic anhydride after separation [3.58].

Only brief reference is made here to the remaining non-C_4-routes to MMA (see Fig. 3.23). The hydroformylation of propene with following oxidation to methacrolein requires more energy than the oxidation of *i*-butene. The same applies to hydrocarboxylation to *i*-butyric acid. Considerable research work is still required for the route via methylacetylene, which can be obtained directly from steam cracker fractions. Ashland, Norsolor, and Röhm are doing R and D on the propene route. The route from ethylene also demands a large amount of energy because it involves hydrocarboxylation to propionic acid and needs also formaldehyde, to be made from methanol. Further, the conversion of propionic acid to formaldehyde in the process hitherto used by Air Reduction is sluggish and requires a complicated recycling procedure [3.111].

3.3.4.4 Isoprene

For a long time no suitable polymerisation procedure could be found for isoprene, known since 1860 as the monomeric building block of natural rubber, which yielded products with properties even remotely approaching the excellence of those of the natural material. Only after intensive basic research, including development of analytical techniques (especially IR-spectroscopy) and discovery of suitable catalysts (organo-metallic Ziegler types) was success achieved. This was in 1954 in the American laboratories of Goodrich with titanium catalysts and, shortly afterwards, at Firestone with lithium catalysts, which enabled the first complete syntheses of poly-(*cis*-1,4-isoprene) as it occurs in nature.

Shell has been producing polyisoprene (IR) since 1960, sold under the name of Cariflex IR (lithium type). Goodyear has been offering its product named Natsyn (titanium type) since 1962. Although the technological breakthrough had already been made at that time, the competition of the natural rubber material was still so strong that the commercial expansion of the synthetic products in the western industrialised nations was rather slow. Their production capacity was 281,000 t/a in 1979. However, IR production in East bloc countries had grown to an estimated amount of 900,000 t/a during the same period [3.123].

IFP, Bayer and Kuraray have developed two-stage procedures for making monomeric isoprene on the C_4-basis. Pure *i*-butene and formaldehyde are used as raw materials. The conversion takes place via dimethyl-*m*-dioxan (DMD):

$$\underset{CH_3}{\overset{CH_3}{>}}C=CH_2 \ + \ 2HCHO \longrightarrow \underset{CH_3}{\overset{CH_3}{>}}C\underset{CH_2-CH_2}{\overset{O-CH_2}{<}}O \longrightarrow \underset{CH_3}{\overset{CH_2=CH}{>}}C=CH_2 + HCHO + H_2O$$

Formaldehyde can be produced by catalytic dehydrogenation or oxydehydrogenation of methanol.

The first reaction step (Prins reaction) takes place in the presence of acid catalysts. IFP and producers in the USSR use a 1 to 1.5% sulfuric acid for this reaction [3.124]. Bayer were able to reduce the difficulties due to corrosion and processing of the waste water arising from the use of the liquid catalyst, by developing an acidic ion exchanger [3.125]. Alongside the intermediate DMD, polyols (Residol-1), mainly dioxan-alcohols and piperylene, are formed, provided that n-butenes are contained in the feed stock. Up to 25% of Residol-1, based on the isoprene product, can be formed [3.124]. The Residol-1, of only limited market potential, is separated in the IFP procedure. In the Bayer process it is transferred into the following decomposition reaction.

A mixture of isoprene, formaldehyde and steam is formed by decomposition of DMD on a fluid bed catalyst (phosphoric acid on a carrier) at 200 to 350 °C. This forms two phases on quenching. The organic phase contains isoprene, i-butene, unconverted DMD and an oily residue (8% Residol-2, based on isoprene) which are separated in three distillation stages. The aqueous phase contains formaldehyde, which, together with i-butene and DMD, are recycled [3.126].

Up to the middle of the 1970s work was performed in Japan on developing one-stage i-butene/formaldehyde processes with solid bed catalysts, whereby TBA can also be used directly instead of i-butene [3.127].

Marathon in the USA developed a new version of this process, in which formaldehyde is first converted into chloromethyl ether which is then reacted with i-butene on TiCl₄:

$$HCHO \ + \ CH_3OH \ + \ HCl \longrightarrow CH_3OCH_2Cl \ + \ H_2O$$

$$\underset{CH_3}{\overset{CH_3}{>}}C=CH_2 + CH_3OCH_2Cl \longrightarrow \underset{CH_3}{\overset{CH_3}{>}}CCl-CH_2CH_2OCH_3 \longrightarrow \underset{CH_3}{\overset{CH_2=CH}{>}}C=CH_2 + CH_3OH + HCl$$

Chlorodimethyl-
propyl ether

After decomposition at 105 to 165 °C the isoprene is extracted with NMP, and chloromethyl ether obtained from the residual mixture by recycling. The yield of isoprene, based on i-butene, is claimed to be about 80% [2.124].

Methylbutenes (isoamylenes) are formed by disproportionation of a mixture containing only i-butene and n-butenes. These can be dehydrogenated to isoprene in a further step [3.128]:

$$\begin{array}{c} CH_3-CH=CH-CH_3 \\ CH_2=CH-CH_2-CH_3 \end{array} + \ 2CH_3-\overset{CH_3}{\underset{|}{C}}=CH_2 \longrightarrow \begin{array}{c} CH_3-CH=\overset{CH_3}{\overset{|}{C}}-CH_3 \\ CH_2=\overset{CH_3}{\underset{|}{C}}-CH_2-CH_3 \end{array} + \ 2CH_3-CH=CH_2$$

Pure propene is yielded as co-product. Reaction conditions and catalysts correspond to those of the propene disproportionation, already described (see Sect. 2.2.4.1).

A pure butene mixture could easily be prepared from raffinate I by the Olex process of UOP (see Sect. 3.1.2.3).

Isoprene can be made from a butene mixture also by using methylal [3.129]:

$$\begin{array}{c} CH_3 \\ | \\ CH_2=C-CH_3 \\ CH_3-CH=CH-CH_3 \end{array} + 2CH_2(OCH_3)_2 \longrightarrow 2CH_2=\overset{\overset{\displaystyle CH_3}{|}}{C}-CH=CH_2 + 4CH_3OH$$

i-Butene and, to some extent, 2-butene are reacted to give isoprene. Under the conditions chosen neither 1-butene nor the butanes are converted.

Despite these many developments, only one commercially operated plant on C_4-basis that is not in the USSR is in use today. This was built by Kuraray in Japan in 1973, with a capacity of 30,000 t/a taking the route via DMD described above.

There are some other isoprene syntheses based, however, on other raw materials [3.130]:

1) The catalytic dehydrogenation of *i*-pentane according to Houdry:

$$2CH_3-\overset{\overset{\displaystyle CH_3}{|}}{CH}-CH_2-CH_3 \longrightarrow CH_2=\overset{\overset{\displaystyle CH_3}{|}}{C}-CH=CH_2 + CH_3-\overset{\overset{\displaystyle CH_3}{|}}{C}=CH-CH_3 + 2H_2$$

                                                                    2-Methyl-2-butene

$$\overset{CH_3}{\underset{CH_3}{>}}C=CH-CH_3 \longrightarrow CH_2=\overset{\overset{\displaystyle CH_3}{|}}{C}-CH=CH_2$$

This illustrates the feedstock flexibility of the Houdry process, which was formerly widely used to dehydrogenate *n*-butane to butadiene (see Sect. 3.2.2). Large amounts of *i*-pentane are available in most refineries equipped with a unit for C_5/C_6-isomerisation. Only in the Soviet Union has the process been technically accomplished.

2) The Dow or Shell catalytic dehydrogenation of isoamylenes:

$$CH_3-CH=\overset{\overset{\displaystyle CH_3}{|}}{C}-CH_3 + CH_2=\overset{\overset{\displaystyle CH_3}{|}}{C}-CH_2-CH_3 \longrightarrow 2\,CH_2=\overset{\overset{\displaystyle CH_3}{|}}{C}-CH=CH_2 + H_2$$

Up to 10% by wt. of isoamylenes are contained in catalytic cracking gasoline. They must be extracted, e. g. with sulfuric acid. Based on this, Shell built isoprene plants in Europe (Pernis, 70,000 t/a), and in the USA (Marietta, OH, 45,000 t/a and Torrance, CA, 18,000 t/a), and also issued a licence to Goodrich-Gulf which led to the construction of a 54,000 t/a plant in Orange, TX.

3) Dimerisation of propene according to Scientific Design-Goodyear:

$$2CH_2=CH-CH_3 \longrightarrow CH_2=\overset{\overset{\displaystyle CH_3}{|}}{C}-CH_2-CH_2-CH_3 \longrightarrow CH_3-\overset{\overset{\displaystyle CH_3}{|}}{C}=CH-CH_2-CH_3 \longrightarrow CH_2=\overset{\overset{\displaystyle CH_3}{|}}{C}-CH=CH_2 + CH_4$$

            2-Methyl-1-pentene   2-Methyl-2-pentene

Up to 1974 Goodyear operated a large scale plant for 50,000 t/a in Beaumont, TX utilizing this three-stage process. It was then destroyed by a disastrous fire and never rebuilt.

4) The reaction of acetone with acetylene, according to Snamprogetti/Anic:

$$CH_3-CO-CH_3 + CH\equiv CH \longrightarrow CH_3-\overset{\overset{\displaystyle OH}{|}}{\underset{\underset{\displaystyle CH_3}{|}}{C}}-C\equiv CH \xrightarrow{+H_2} CH_3-\overset{\overset{\displaystyle OH}{|}}{\underset{\underset{\displaystyle CH_3}{|}}{C}}-CH=CH_2 \xrightarrow{-H_2O} CH_2=\overset{\overset{\displaystyle CH_3}{|}}{C}-CH=CH_2$$

The Italian Anic started production in 1973 at a plant in Ravenna, based on this three-stage procedure; it continued until the beginning of the 1980s [3.131].

Selective extraction of isoprene from the C_5-fraction from steam crackers is the dominating process in the western world today. About 15 to 20% of this fraction, based on the ethylene amount, is yielded as co-product, the higher figure being attained under less severe cracking conditions. The fraction contains 10 to 20% of the useful diolefines, isoprene and cyclopentadiene. For the third diolefine, piperylene, there are at present insufficient market outlets. The C_5-extraction procedures are largely the same as for butadiene extractions. The same solvents can be employed (see Sect. 3.1.1). The extraction capacity for isoprene in the western world amounted to 220,000 t/a in 1976.

3.3.4.5 Specialities

Most C_4-specialities are derivatives of pure *i*-butene. This complex field is dealt with here in tabular form, without details of the process techniques (Table 3.11). Data concerning production are not readily available.

3.3.5 Chemical Secondary Products from Pure *n*-Butenes

Pure 1-butene is necessary for all *n*-butene polyolefines. Its specifications are shown in Table 1.5 in Sect. 1.4. To prepare butene oxide a mixture must be used which contains 99.9% of the two *n*-butene isomers.

3.3.5.1 Poly-1-butene

Pure 1-butene can be polymerised on special Ziegler-Natta catalysts to a product which has a far higher molecular weight (0.8 to 3×10^6) than the other polyolefines. These catalysts favour formation of highly isotactic polymers. The structure corresponds to that of polypropene in which the branched methyl groups have been replaced by ethyl groups. Production is carried out batchwise in cooled tubular autoclaves at temperatures of 60 to 90 °C and pressures of 8 to 15 bar, so as to keep the 1-butene in the liquid phase. Certain polypropene plants can be easily adapted for this process. The degree of polymerisation is controlled by metering the added hydrogen.

The polymerisation is succeeded by a washing stage to remove catalyst residues, and a separating stage in which about 50 kg atactic poly-1-butene (PB) is removed per ton of product [3.132]. The polymer product is then dried and compounded. The atactic by-product can also be marketed, just as the atactic polypropene. During the 1970s, Hüls produced 12,000 t/a PB in a pilot plant in the FRG [3.133]. Since then, Shell in the USA is the only producer, with a plant in Taft, LA (22,000 t/a), which belonged to Witco Chemical until 1977 [3.134]. Shell uses as raw material pure 1-butene from their SHOP plant in Geismar, LA. Neste Oy and Idemitsu Petrochemical started together in 1985 to develop a new poly-1-butene process. They planned to start production experimentally within three years [3.135].

Table 3.11. *i*-Butene specialities and their fields of use, classified according to reaction type

Reaction type	Reagent	Products	Types of product, fields of use
Chlorination and conversion to, with ring cleavage	Cl_2	*tert*-butyl chloride (= methallyl chloride) chlorohydrin, dichlorohydrin	pharmaceuticals, solvents
		β-methylepichlorohydrin	epoxy resins with bisphenol A (epiclones of Dai Nippon Inc.)
Sulfochlorination	Cl_2/SO_2	methallyl sulfonate	surface-active agents
Formation of metal complexes	Al	di- or tri-butylaluminium	catalysts
Reaction with	acetic acid	*tert*-butyl acetate	gasoline additive (synerg. improvement of octane number)
Reaction with	NH_3	*tert*-butylamine	intermediate for lubricant additives, rubber auxiliaries, pharmaceuticals, plant protection agents
Reaction with	H_2S	*tert*-butanethiol	odoriser
Hydroformylation and hydrogenation or oxidation	CO, H_2 H_2 O_2	*i*-valerianaldehyde 3-methyl-1-butanol *i*-valerianic acid	solvent stabilisers
Hydrocarboxylation	CO, H_2O	trimethylacetic acid (= neopentanoic or pivalic acid)	polymerisation initiator, drying oil component for varnishes
Resin formation	MA	copolymers of MA and *i*-butene	polyesters
Alkylation of aromatics	phenol	4-*tert*-butylphenol	phenolic resins, plasticisers for cellulose acetate
		2,6-di-*tert*-butylphenol 2-*tert*-butylphenol	gasoline additive antioxidant
	toluene	4-*tert*-butyltoluene	alkyd resins, polyesters, perfumes
	p-cresol	2,6-di-*tert*-butylcresol (TBC) [= butylated hydroxytoluene (BHT)]	antioxidant, polymsn. inhibitor
	pyrocatechol	4-*tert*-butylcatechol	antioxidant, polymsn. inhibitor
	p-hydroxyanisole	2- or 3-*tert*-butyl-4-hydroxyanisole (BHA)	food preservative

3.3.5.2 HDPE-Comonomer and LLDPE

In the production of high density polyethylene (HDPE) up to 4% 1-butene has been added as comonomer for a long time. This is usually done by those HDPE producers whose low pressure polymerisation plants were erected in the middle of the 1970s and based on the suspension processes of Ziegler, Phillips (Marlex) and Standard Oil. The copolymerisation interferes with the formation of the crystalline molecular structure and softer products of lower density are obtained. Larger proportions of comonomer can lead to swelling. This causes yield losses and hence the density of suspension-HDPE has a lower limit of about 0.930 g/cm^3.

The linear low density polyethylene (LLDPE) with a density down to 0.915 g/cm^3 is a comparatively new product. It is generally an HDPE containing a comonomer amount of about 8 to 12% by wt. of 1-butene. Gas phase processes above all are suitable for its preparation. The Unipol fluid bed process, commercialised by UCC in 1977 is regarded as the market leader within this new generation of processes. Several licensed plants have been erected around the world [3.136]. BP built a gas phase fluid bed LLDPE plant utilizing to its own process in Lavera/France in 1985, and issued a license to USI [3.137]. ROW is running a 20,000 t/a gas phase experimental plant for LLDPE in Wesseling.

The solution polymerisation processes of Dow, Du Pont Canada and Mitsui constitute a second group of low pressure LLDPE processes. Their technical operation is more elaborate than the gas phase processes. Additional equipment is needed for separating solvent and also, as a rule, oligomers and waxes from the product. A limit exists for the preparation of higher molecular materials because of the broad molecular weight distribution yielded by the increased viscosity of the polymer solution. However, within their smaller range of product types compared to the gas phase processes, they permit a more flexible changeover of type during production.

Ruhrchemie, Oberhausen, a production plant of Hoechst, has had a continuous high pressure LLDPE process since 1983, based on their own research. It was tested in a 15,000 t/a line of a revamped LDPE plant [3.138]. It is possible to produce LDPE, MDPE or HPDE by slight modifications of the catalyst system but the product range is limited in comparison to the low pressure polymerisation. Refitting existing LDPE plants is said to cost only about 10 to 15% of the investment in a new installation. CdF-Chimie in France is producing high pressure LLDPE in rebuilt LDPE plants with a new catalyst system in a discontinuous autoclave process. Additional LLDPE plants have been built or rebuilt in France, Spain, and Japan, based on licenses from CdF-Chimie. Enichem in Italy also carried out experiments on high pressure processes, using catalysts developed by El Paso Olefins in collaboration with Montedison. Despite this, Enichem decided to build a new low pressure LLDPE plant based on the established process of Du Pont Canada [3.137].

Table 3.12 shows product capacities for LLDPE. Some producers also use higher olefines as comonomers, e. g., 1-octene and BP, 4-methyl-1-pentene. 1-Octene, in particular, could be obtained in greater quantities from raffinate II (Octol process, Sect. 3.3.3.1). It can be seen that LLDPE is largely a commodity brought on to the market by American and Arabian companies.

However, BP have the intention of building additional LLDPE plants in Europe and have recently named Grangemouth in Great Britain and Cologne-Worringen in the FRG (EC in a joint venture with Bayer) as possible sites for capacities of about 100,000 t/a in each case [3.140].

3.3.5.3 1,2-Butene Oxide

1,2-Butene oxide is yielded with 15% amount as a by-product in the synthesis of 2,3-butanediol from raffinate II (see Sect. 3.3.3.6). Dow Chemical is the only producer of 1,2-butene oxide in the USA. It is obtained, like propene oxide, via the chlorohydrin

Table 3.12. World capacities for LLDPE in 1985 and additional capacities up to the end of 1987 from plants under construction [3.142]

	Company	Site	Process	Capacity 1985	by 1987
N. America					2440 + 300
Canada	Dow	Ft. Saskatchewan, AL	NL		+ 120
	Du Pont	Montreal, PQ	NL	235	
	Esso	Sarnia, ON	NG	135	
	Novacor	Joffre, AL	NG	270	
	UCC	Sarnia, ON	NG	115	
USA	Dow	Freeport, TX	NL	135	+ 30
		Plaquemine, LA	NL	235	+ 50
	Exxon	Mt. Belview, TX	NG	320	
	Mobil	Beaumont, TX	NG	160	
	Norchem	Joliet, IL	NG	110	
	UCC	Seadrift, TX	NG	360	
		Taft, LA	NG	365	
	USI	Port Arthur, TX	NG		+ 100
S. America					120
Argentina	Polisur	Bahia Blanca	NG	120	
W. Europe					600 + 300
France	Atochem	Balan	H	35	
	BP Chimie	Lavera	NG	60[4]	
	CdF Chimie	Dunkerque	H	100	
		Lillebonne	H	55	
Italy	Enichem	Priolo	NL		+ 140
Netherlands	Dow	Terneuzen	NL	90	
	DSM	Geleen	NL	40	+ 60
Austria	Petrochemie Danubia	Schwechat	NG		+ 80
Sweden	Neste Oy	Stenungsund	NG	150	
Spain	Alcudia	Puertollano	H	20	+ 20
	Dow	Tarragona	NL	50[5]	
E. Europe					0
Remaining world					1060
Australia	Altona	Altona	NG	55[6]	
Japan	MYKK	Kashima	NG	75	
	Nippon/UCC	Kawasaki	NG	75	
	Mitsui	Chiba	NL		
	Sumitomo	Niihama	H	190	
	Toyo Soda	Yokkaichi	H		
Saudi Arabia	Al Jubail Petrochem.[1]	Al Jubail	NG	260	
	Eastern Petrochem.[2]	Al Jubail	NG	130	
	Saudi Yanbu[3]	Yanbu	NG	200	
S. Africa	AECI	Sasolburg	NG	75	
Total					4220 + 600

H: High pressure   NG: Low pressure, gas phase   NL: Low pressure solution

[1] 50% Exxon, 50% Sabic

[2] 50% MYKK, 50% Sabic

[3] 50% Mobil Oil, 50% Sabic

[4] estimated 50% from new 120,000 t/a HDPE/LLDPE plant

[5] estimated 50% from new 100,000 t/a HDPE/LLDPE plant

[6] estimated 50% from the 110,000 t/a plants belonging to UCC up to 1983

route. A feed mixture is used which contains 99.9% by wt. of *n*-butene. At the beginning of the 1970s about 5000 t/a 1,2-butene oxide were still being made but production has fallen to less than 500 t/a in the meantime. The product is used as the principal component of a stabilising system for chlorinated hydrocarbon solvents. The use of such solvents (e. g. for degreasing of metals) also been reduced markedly as a result of new legal measures concerning recycling and replacing them by other solvents.

3.4 Processes of Butadiene Conversion

3.4.1 Butadiene Polymers and Copolymers

The polymers and copolymers of butadiene are important synthetic rubbers. Polybutadiene and polychloroprene belong to the polymers of butadiene and its secondary products. The most important copolymers are styrene-butadiene-rubbers. Nitrile rubbers have also to be mentioned in this context. Butadiene therefore appears on the plastics scene, as a component of polystyrene, styrene co- and terpolymers, and PVC for improving impact resistance. Data on the capacities for rubbers and latices are published constantly by the International Institute of Synthetic Rubber Producers (IISRP).

Tables 3.13 and 3.14 give a survey of the capacity and consumption data of the most important synthetic rubbers. About three-quarters of the quantitatively most important synthetic rubbers, styrene-butadiene-rubber and polybutadiene, are used in the tires sector. These products are therefore particularly dependent on the market developments in the automobile industry.

3.4.1.1 Polybutadiene

Polybutadiene (BR) is a stereospecific polymer, the research on which was closely related with the Ziegler-Natta-catalysts in the 1950s. These catalysts impart to the polymer a differentiated pattern of properties. Table 3.15 depicts the various catalysts and the products obtained.

Isotactic 1,2-polybutadiene and 1,4-*trans*-polybutadiene are high molecular, crystalline plastomers, whereas the high-molecular 1,4-*cis*-polybutadiene is an elastomer. Low molecular 1,4-*cis*-polybutadiene is a liquid of low to high viscosity.

All BR types are prepared by polymerisation in benzene solution. Butadiene, benzene, the catalysts and a polymerisation modifier are brought to reaction in a tank

Table 3.15. Catalysts and products of butadiene polymerisation [3.144]

Catalyst	Co-catalyst	Product	Mol. weight
Triethylaluminum	Cr- or Co-compound	isotactic 1,2-polybutadiene	>100,000
Triethylaluminum	Ti-compound	1,4-*cis*-polybutadiene	>100,000
Diethylaluminum	Co-compound	1,4-*cis*-polybutadiene	>100,000
Ethylaluminum-sesquichloride	V-compound	1,4-*trans*-polybutadiene	>100,000
Ethylaluminum-sesquichloride	Ni-compound	liquid 1,4-*cis*-polybutadiene	500 to 30,000

Table 3.13. Synthetic rubber capacities, classified according to important regions and types, worldwide for 1983 and 1985 (in 1000 t/a) [3.142]

Type Region	Styrene-Butadiene (SBR)		Poly-butadiene (BR)		Poly-isoprene (IR)		Ethylene-propene (EPM, EPDM)		Butyl (IIR)		Poly-chloroprene (CR)		Nitrile (NBR)		Total synthetic rubber	
Year	1983	1985	1983	1985	1983	1985	1983	1985	1983	1985	1983	1985	1983	1985	1983	1985
World total	6696	6936	1584	1664	1224	1224	502	540	693	696	556	556	508	508	11763	12124
Countries with planned economy	1550	1765	245	290	935	935	–	–	110	110	125	125	88	88	3053	3313
Countries with free market economy	5146	5171	1339	1374	289	289	502	540	583	586	431	431	420	420	8710	8811
of which USA	1487	1487	413	434	60	60	243	251	219	219	213	213	119	119	2754	2783
Canada	142	142	63	63	–	–	–	–	120	120	–	–	20	20	345	345
Total N. America	1629	1629	476	497	60	60	243	251	339	339	213	213	139	139	3099	3128
Total Latin America	404	419	84	94	12	12	–	–	–	–	–	–	11	11	511	536
Total Asia, Africa, Australia	1250	1250	371	371	142	142	75	85	60	60	85	85	92	93	2075	2086
Total W. Europe	1863	1873	408	412	75	75	184	204	184	187	133	133	178	178	3025	3062
Total EEC-Countries	1679	1679	374	378	75	75	184	204	184	187	133	133	176	176	2805	2832
of which FRG	410	410	70	70	–	–	27	27	–	–	60	60	47	47	614	614
France	357	357	164	164	35	35	42	42	47	47	40	40	37	37	722	722
United Kingdom	307	307	60	60	–	–	15	15	52	55	33	33	15	15	482	485
Italy	350	350	70	70	–	–	50	50	–	–	–	–	47	47	517	517
Benelux	255	255	10	14	40	40	50	70	85	85	–	–	30	30	470	494

Table 3.14. Rubber consumption in W. Europe and N. America (in 1000 t/a) [3.143]

Type of rubber / Year	W. European consumption					N. American consumption				
	1983 (eff)	1984	1985	1986	1984 % tires²	1983 (eff)	1984	1985	1986	1984 % as tires
Styrene-butadiene rubber (dry)	594	603	615	649	66	873	930	929	969	73,4
Styrene-butadiene-latex	109	109	110	112	5	104	107	108	114	6,3
Polybutadiene	222	232	237	260	75	386	430	433	464	77,2
Ethylene-propene rubber	133	140	146	165	–	148	184	193	227	8,2
Nitrile rubber (dry and latex)	73	77	78	80	–	60	70	72	78	–
Polychloroprene	81	80	79	77	–	89	95	97	104	–
Other synthetic rubbers¹	192	200	208	237	64,5	352	393	404	431	46,8
Total synthetic rubbers	1404	1441	1473	1580	47,7	2012	2209	2236	2389	55,3
Natural rubber	826	840	850	890	68	760	839	834	876	75,1
Total new rubber	2230	2281	2323	2470	55,2	2772	3048	3070	3265	60,7
% of synthetic rubber	63,0	63,2	63,4	64,0	–	72,6	72,5	72,8	73,2	–
Carboxylated styrene-butadiene (latex)	255	273	275	290	–	337	359	367	403	–

¹ Including butyl and polyisoprene
² including tires and tire products

cooled by brine. After polymerisation the catalyst components are washed out from the solution with water, and this wash-water is freed from benzene by steam. The washed, low-molecular polymer solution is concentrated in a film evaporator. Benzene recovered here can be reprocessed and recycled.

Important producers of BR, together with the associated trade names are: American Synthetic Rubber (Cisdene), Firestone (Diene), General Tire and Rubber (Duragen), Goodyear Tire and Rubber (Budene, Plioflex, Tufsyn), Hüls (Buna CB), International Synthetic Rubber (Intene), and Shell (Cariflex BR).

In the FRG Bunawerke Hüls (BWH) has a total capacity for 70,000 t/a cis-BR-rubber and 10,000 t/a liquid polybutadiene (Polyöl Hüls).

3.4.1.2 Polychloroprene

It has been possible to produce polychloroprene (CR) on the basis of butadiene since the middle of the 1950s. The fundamental development work was done by Distillers Co., later acquired by BP chemicals [3.145]:

1) Chlorination of butadiene in the gas phase at 300 °C

$$CH_2=CH-CH=CH_2 + Cl_2 \nearrow \begin{array}{l} CH_2=CH-CHCl-CH_2Cl \\ 38\% \; 3,4\text{-Dichloro-1-butene} \\ \searrow \; CHCl-CH=CH-CH_2Cl \\ 62\% \; 1,4\text{-Dichloro-2-butene} \end{array}$$

2) Isomerisation of 1,4-dichloro-2-butene in the liquid phase at 130 to 145 °C, on CuCl$_2$ (unless it is directly used for preparing adiponitrile, see Sect. 3.4.3, or THF/1,4-butanediol, see Sect. 3.4.4)

$$CH_2Cl-CH=CH-CH_2Cl \longrightarrow CH_2=CH-CHCl-CH_2Cl$$

3) Dehydrochlorination of 3,4-dichloro-1-butene with NaOH at 80 °C

$$CH_2=CH-CHCl-CH_2Cl + NaOH \longrightarrow CH_2=CCl-CH=CH_2 + H_2O + NaCl$$
$$\text{Chloroprene}$$

4) Polymerisation of the chloroprene with persulfate or a redox system as initiator at 10 to 40 °C

$$n \; CH_2=CCl-CH=CH_2 \longrightarrow \{CH_2-CCl=CH-CH_2\}_n$$

A flow chart of the process is given in Fig. 3.27.

The raw materials, chlorine and butadiene, must be dry and almost free of oxygen. Side reactions, such as chlorine substitution, excess chlorination or thermal dehydrochlorination lead to the formation of HCl and of chlorinated hydrocarbons of low and high boiling points as residual products. Volatile and non-volatile C$_4$–C$_8$ products and tars are formed, so that the chlorine content lies between 40 and 60% [3.146]. If the optimum process conditions are achieved the formation of by-products can be kept within narrow limits. Butadiene and HCl are withdrawn at the head of the degasing column. Aqueous hydrochloric acid is separated from this stream in the HCl-washer, whereas butadiene is recycled. The involatile residue is separated as the bottom product during the isomerisation and purification of the 3,4-dichloro-1-butene. After dehydrochlorination the chloroprene is obtained in pure form by

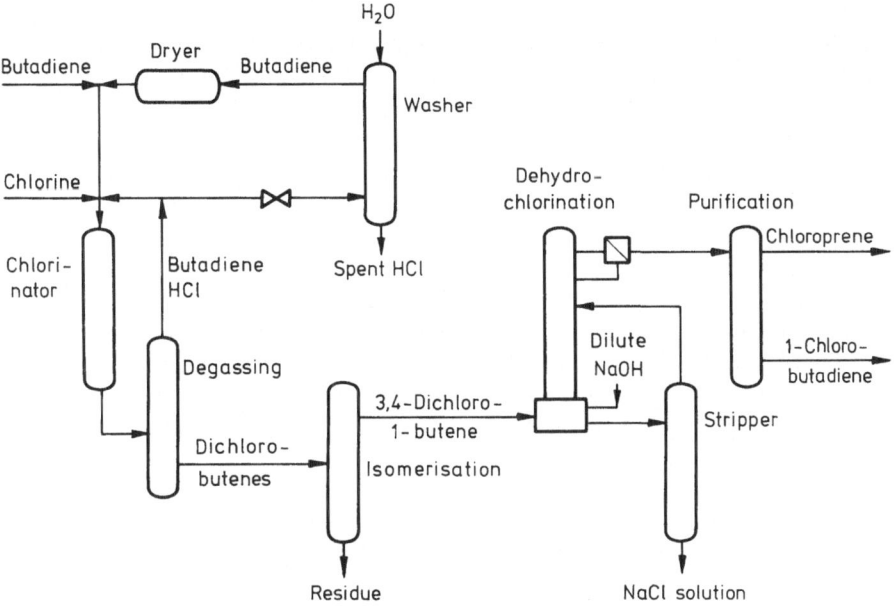

Fig. 3.27. Chloroprene production from butadiene and chlorine [3.145]

separating it from unreacted 3,4-dichloro-1-butene, which is returned to the reaction, and from 1-chlorobutadiene, formed as a by-product.

During dehydrochlorination about 5 t of a 15% solution of common salt is formed per t of chloroprene. The solution also contains excess sodium hydroxide, chlorinated hydrocarbons and polymerisates. The disposal of this waste product has so far presented problems which have not yet been satisfactorily solved.

Suitable raw materials are pure butadiene and also C_4-cuts which contain the n-butenes and n-butane in addition to butadiene. Hydrogen chloride is formed as a side product during chlorination of the C_4-hydrocarbons which are more saturated than butadiene. This must be utilized, e. g. in vinyl chloride production from hydrochlorination of acetylene or oxychlorination of ethylene.

$$\left.\begin{array}{l} CH_3-CH=CH-CH_3 \\ CH_2=CH-CH_2-CH_3 \end{array}\right\} + Cl_2 \longrightarrow \left\{\begin{array}{ll} CH_2Cl-CH=CH-CH_3 & + \;\; HCl \\ CH_2=CH-CHCl-CH_3 & + \;\; HCl \\ CH_2Cl-CH=CHCl-CH_3 & \end{array}\right.$$

$$CH_3-CH_2-CH_2-CH_3 \xrightarrow{\;+Cl_2\;} CH_2Cl-CH_2-CH_2-CH_3 + \;\; HCl$$

The n-monochlorobutenes and n-chlorobutanes can be separated from the dichlorobutenes derived from butadiene and then decomposed in a tubular reactor at 600 °C into butadiene or n-butenes and HCl. The butadiene and the butenes are subsequently fed back into the chlorination where they are progressively converted into the desired dichlorobutenes.

The emulsion polymerisation of the chloroprene is performed discontinuously in brine-cooled agitator vessels. Added are an emulsifier (soap solution), surface-active agents, an initator, and either sulfur or thiols (formerly mercaptans) as modifiers. When the densitiy of the emulsion (latex) has attained the desired value, the polymerisation is halted by adding a stopper-solution. The sulfur-modified types are finally treated with thiuram disulfide (peptisation). The latex is stripped with steam to remove monomer and then coagulated (i. e. the influence of the emulsifier is eliminated by adding acids or salts) which is carried out on cooling rollers. The formed rubber film is washed, dried and cut into chips [3.147]. The wash water requires two treatments. Its residual latex content must be specially coagulated, separated by filtration, and disposed of. The chlorinated hydrocarbons in the waste water must be degraded biologically because incineration would consume too much energy and also cause HCl emission.

Distugil (50% BP, 25% Plastugil, 15% Rhônes-Alpes and 10% E. R. A. P.) started production in 1966 at the first commercial plant for preparing chloroprene from butadiene and chlorine, in Champagnier, France. The plant has a present capacity of 40,000 t/a. In the meantime other plants of this type have been built under BP-license: 1975 by Bayer in Dormagen (60,000 t/a today), by Toyo Soda in Shin-Nanyo, Japan (12,000 t/a at present), and 1970 by Petro-Tex in Houston, TX, a plant taken over by Denka in 1977 and expanded (34,000 t/a today). Du Pont, the world's most important producer of chloroprene, has its own experience and know-how of the butadiene route. The company originally utilized the acetylene route in plants in the USA and Northern Ireland until the late 1970s.

In the first stage of the acetylene process, the acetylene is dimerised to monovinyl-acetylene at about 70 °C on contact solutions with copper(I)-chloride and alkali or ammonium salt. The monovinylacetylene is then converted to 4-chloro-1,2-butadiene by 1,4-addition of HCl; this is rearranged to chloroprene in the presence of the catalyst solution containing copper(I)-chloride:

$$2\,CH{\equiv}CH \longrightarrow CH_2{=}CH{-}C{\equiv}CH \xrightarrow{+HCl} CH_2Cl{-}CH{=}C{=}CH_2 \longrightarrow CH_2{=}CCl{-}CH{=}CH_2$$

During the first stage, the formation of monovinylacetylene (75 to 95% yield), by-products are formed, notably divinylacetylene, acetaldehyde and vinyl chloride. The side product 1,3-dichloro-2-butene is formed by the secondary reaction of HCl addition to the chloroprene:

$$CH_2{=}CCl{-}CH{=}CH_2 \ + \ HCl \longrightarrow CH_3{-}CCl{=}CH{-}CH_2Cl$$

Utilization of this compound is possible, i. e. processing it to 2,3-dichloro-butadiene which can be employed as a comonomer of chloroprene. There is altogether a yield of chlorinated hydrocarbon residues as in the butadiene process. Chlorine contents of 40 to 60% are present in the residues [3.146].

The explosive nature of vinylacetylene demanded extensive security precautions so that the acetylene process needed greater investment costs than the butadiene route. Estimates by experts were of about one-third more [3.148]. The high costs of energy and raw materials were also disadvantageous. On the butadiene basis the energy consumption is 22 MJ/kg chloroprene, markedly less than the 27.7 MJ/kg of the acetylene route. If the energy contents of the raw materials is included, the butadiene process with 141 MJ/kg is far more favourable than the acetylene route with 230 MJ/kg [3.149].

Du Pont carried out extensive chemical engineering improvements to the acety-lene process, aimed at continuous operation without intermediate storage of the vinylacetylene. However, after explosion damage at the Louisville factory, the firm went over completely to the butadiene route. Du Pont possesses a CR capacity of 179,000 t/a in the USA at its works in Louisville, KY and Laplace, LA; and over 35,000 t/a in Londonderry, Northern Ireland. It has also awarded a license to the Japanese Showa Neoprene, which is able to produce 20,000 t/a CR at Kawasaki.

Polychloroprene (CR) is used instead of natural rubber and styrene-butadiene-rubber when its special properties are desired, namely those of resistance to weather corrosion and ageing; stability in the presence of ozone, chemicals, and oils; and high flame resistance. The products have been used for many technical rubber articles [3.150].

Customary trade names for polychloroprene are: Neoprene (Du Pont, Showa), Baypren (Bayer), Butachlor (Distugil), Skyprene (Toyo Soda), and Denka Chloroprene (Denka).

3.4.1.3 Styrene-butadiene Rubber

Most of the butadiene is used for the production of styrene-butadiene-rubber (SBR), the most important synthetic rubber. It is produced by co-polymerisation with about 25% styrene, as solid rubber in bales or as a dispersion (latex). The products are referred to as resins when larger styrene amounts are used:

Butadiene content 75%		40%	
(styrene content = difference from 100%)			
Rubber	SBR	Resin	HSBR
Latex	SBL	Latex	HSBL

Solution and emulsion-polymerisation are employed for manufacture. Solution-polymerisation, now only rarely used, e. g. by Firestone, in THF, is initiated by lithium alkyls and has an anionic mechanism. The dominating emulsion-polymerisation with a radical mechanism, is almost entirely performed at low temperatures (cold rubber production), using a redoxsystem. Emulsifiers are added to increase the particle number in the emulsion and thereby to accelerate the polymerisation reaction. This is achieved at the expense of formation of very small latex particles which, however, yield rubber flakes by coagulation but for a long time were not suitable for producing latex with satisfactory qualities. Suitable latex particles could be produced only by abandoning the use of emulsifiers to a large extent. This increased costs because of inferior space/time yields. Not until the beginning of the seventies was success attained in increasing particle size, namely by techniques applied after the emulsion polymerisation, such as freeze-, pressure- or salt-agglomeration. The process of salt agglomeration, sensitized by ethoxylate, used by Hüls since 1973, is more economic in terms of energy than the other techniques.

The production capacities for SBR and SBL in the FRG are given in Table 3.16. Most of the SBL is made as a carboxylated product.

SBR is used chiefly for tires and numerous rubber parts. SBL finds its principal use in dispersion paints and carpet backings.

Table 3.16. Capacities for SBR and SBL in the FRG in 1984

Company	Site	Capacities in 1000 t/a		
		SBR	SBL	carboxylated SBL
BWH (Hüls, Bayer)	Marl	180	–	–
Hüls	Marl	30	62	35
Bayer	Leverkusen	–	–	35
BASF	Ludwigshafen	–	–	10
Dow	Rheinmünster	–	–	40
Synthomer	Langelsheim	–	–	28

3.4.1.4 Nitrile Rubber

Nitrile rubber (NBR) is a copolymer of butadiene and acrylonitrile. It is likewise available as emulsion polymerisate, solid material or latex. The weight proportions butadiene/acrylonitrile range from 55/45 to 82/18. The most important property influencing its utilization is its stability towards hydrocarbons (gasoline, mineral oil). It was first made in the 1930s, in the Leverkusen works of I G Farben. Bayer is the sole producer of NBR in the FRG, with a capacity of 27,000 t/a, at Leverkusen. The trade name is Perbunan N. NB-latex is produced in the FRG by BASF (2000 t/a), Bayer (5000 t/a), and Synthomer (3000 t/a).

3.4.1.5 Acrylonitrile-butadiene-styrene Polymers

Acrylonitrile-butadiene-styrene polymers (ABS) are terpolymers containing about 15% acrylonitrile, 20 to 35% butadiene and 50 to 65% styrene. These thermoplastic products are polymerised either in emulsion or suspension. The α-methylstyrene, yielded as co-product during the phenol synthesis using the cumene-process, is a reactive comonomer sometimes used instead of acrylonitrile. Some producers use MMA as a substitute for the acrylonitrile. The properties of the product thus obtained (MBS) are very similar to those of ABS. It is therefore included in production statistics under ABS. In the FRG, Bayer in Leverkusen have an ABS capacity of 100,000 t/a and BASF in Ludwigshafen one of 60,000 t/a. Rohm and Haas produce MBS using the substitute MMA, under the trade name Cyrolite. It is employed mainly as a component for increasing impact-resistance of UV-stable PVC. A higher content of butadiene increases the impact resistance, whereas more acrylonitrile or α-methylstyrene improves the heat stability. ABS is used to manufacture many valuable technical parts and is also suitable for surface improvement by galvanisation. Specialities are the alloys ABS/polycarbonate (Bayblend of Bayer) and the alloys ABS/PVC, ABS/polysulfone and ABS/polyurethane, all made only in the USA. Table 3.17 gives information about the amounts of styrene copolymerisates produced and used, where the styrene proportion is between 85 and 90%.

Table 3.17. Capacities, production and consumption of styrene copolymerisates in the Western World in 1986 [3.159]

Region	Capacity	Production in 1000 t/a	Consumption
Western Europe	750	580	485
North America	950	580	565
Japan	670	550	490
Rest of Asia	350	220	245
Rest of the world	100	50	75
Total	2800	1980	1860

3.4.1.6 Ethylene-propene-diene Polymers

Unsaturated ethylene-propene rubber (EPDM, EP-rubber, EPTR) contains about 5 to 10% diene as a third component, such dienes include ethylidenenorbornene (EN), dicyclopentadiene (DCP), 1,4-hexadiene (HX) and 1,5-cyclooctadiene (COD). They are butadiene derivatives, with the exception of DCP which is a pure C_5-dimer.

EN is the product of Diels-Alder addition of cyclopentadiene and butadiene [3.152]:

Cyclopentadiene is still available in sufficient amounts as a co-product in the isoprene extraction from steam cracker C_5-cuts; it can, however, be prepared also from butadiene and formaldehyde (Prins reaction). HX, which up to now has only been used rarely as the tercomponent, is a coordination catalysed addition product of butadiene and ethylene [3.153]. The preparation of COD is treated in Sect. 3.4.3.

In a joint investigation with participation from Hoechst, Hüls, Copolymer and Hercules Powder, a large number of dienes were tested at the beginning of the 1970s with EN having the best overall rating. Since then, EN has been used predominantly about 85% of the whole tercomponent consumption.

In the Hüls EPDM process, perchlorocrotonic acid esters are used to activate the Ziegler-catalyst system of vanadium oxytrichloride/ethyl aluminum sesquichloride. This enables the amount of vanadium compounds in the reaction mixture to be reduced and purification to be simplified [3.154]. EPDM can be cross-linked just like other rubbers. Hüls in the FRG has been producing EN-containing EPDM batchwise since 1972 under the trade name of Buna AP. The 27,000 t/a plant is suitable also for the alternative production of ethylene-propene copolymers (EPM, EP rubber). EPDM is characterised by resistance to weathering, low temperatures, and ageing. It has been blended recently with other rubbers for the manufacture of tires.

3.4.1.7 Rubber and Latex Specialities Containing Butadiene

By modifying the basic rubber types without introducing new monomers, many firms have been attempting to achieve improvements in technical applications, especially to gain advantages in processing. This has led to new types of synthetic rubber which are in fact specialities but are sometimes already produced in considerable quantities. All the products mentioned here are statistically included in the capacity data of each basic type previously described.

The so-called telechelics, based on BR, SBR and NBR, belong to the rubber specialities. Their olefinic end-groups are saturated to leave terminal hydroxyl, carbonyl or halogen groups. About 12,000 t/a of the BR-capacity of Hüls are reserved for preparing these telechelics. Arco is producing under the brand name of Poly bd a solid BR with terminal hydroxyl groups, in Channelview, TX and, since 1986, also in Botlek, and in a joint venture with Idemitsu in Tokuyama. Phillips produces BR with terminal hydroxyl groups (Butarez CTL) in liquid form, in Borger, TX. Its principal use is as a binder for solid propellants. Thermoplastic, elastomeric polyurethanes can be made from the hydroxyltelechelics by reaction with diisocyanates.

The SBS-triblock polymer also has interesting technical applications. The interior butadiene block is soft and imparts rubberlike properties. Shell has been producing this thermoplastic elastomer (30,000 t/a Cariflex TR) at ROW, Wesseling since 1973. A photograph of this plant is shown in Fig. 3.28 — other Shell plants for this product are located in Berre (30,000 t/a) and in Marietta (125,000 t/a). Some other companies manufacture similar products — Hüls, for example, is making an HSBR-resin with approx. 50% styrene content under the name of Duranit (28,000 t/a).

Fig. 3.28. ROW plant for the production of the thermoplastic rubber Cariflex TR at Wesseling

Attempts have been made for about 10 years to market as a new type of SBR/BR a ready-made free-flowing powdered mixture of rubber and filling materials (master batches). This would be more easily handled in storage, transport, and processing [3.154].

In the meantime more latex is produced as speciality types than as standard types. Specialities account for more than 50% at Hüls, the world's largest producer of latex — carboxylated SBL being most prominent product (see Table 3.16 concerning capacities). Latex specialities are adaptable to the various processing techniques (dipping, coating, pasting, foaming).

3.4.2 Styrene

Butadiene dimerises at 200 °C and, without a catalyst, to 4-vinyl-1-cyclohexene (Diels-Alder reaction):

The reaction occurs at temperatures as low as 10 to 40 °C when complex catalysts and other agents are added. The yield of 4-vinyl-1-cyclohexene is almost quantitative. The intermediate can be oxidised to styrene in a succeeding stage, with air or oxygen at 140 to 250 °C using a palladium catalyst on a carrier.

Benzene is now and will in future not to be used in gasoline, which means that the supply will increase and hence the market price will be depressed. The classical routes to styrene via ethyl benzene is thus likely to remain dominant. These routes include notably the dehydrogenation of ethyl benzene to styrene, and the oxidation of propene with ethyl benzene to propene oxide and styrene as co-product (Halcon, Shell SMPO).

3.4.3 Polyamides

Considerable development work was devoted to the syntheses from butadiene as the base chemical for making polyamides. Adipic acid and hexamethylenediamine give AH salt from which polyamide-6,6 (nylon) is produced, and can now be prepared from butadiene.

In the BASF new adipic acid process, butadiene reacts with carbon monoxide and methanol in the first step, at 120 °C and 600 bar in the presence of cobalt carbonyl and aromatic bases. The reaction gives methyl 3-pentenoate in 80% yield [3.155]. This monoester is separated and converted to dimethyl adipate at 185 °C and 30 bar. This ester can be hydrolyzed to adipic acid:

$$CH_2=CH-CH=CH_2 \xrightarrow{+CO+CH_3OH} CH_3-CH=CH-CH_2-COOCH_3 \xrightarrow{+CO+CH_3OH}$$
$$CH_3OOC-CH_2-CH_2-CH_2-CH_2-COOCH_3 \xrightarrow[-2CH_3OH]{+2H_2O} HOOC-CH_2-CH_2-CH_2-CH_2-COOH$$

There are about equal amounts of n-valerianic and α-methylglutaric acids as by-products.

A C_4-fraction containing about 44% butadiene, from steam crackers, is also suitable as starting material. The butenes, butanes, 1,2-butadiene and C_4-acetylenes present do not interfere in the reaction. Methanol is recovered after the hydrolysis and recycled. A large scale plant producing 60,000 t/a adipic acid with this process was said to be going into operation at BASF in Ludwigshafen at the beginning of 1988.

Prior to this adipic acid was made all over the world mostly by oxidation of cyclohexanol/cyclohexanone mixtures with nitric acid, the mixtures being derived from benzene. World capacity in 1983 was 2016 thousand tons.

For a long time the synthesis of hexamethylenediamine (HMDA) has been based on chlorinating butadiene and adding hydrocyanic acid (Du Pont process). Chlorination of butadiene gives a mixture of two dichlorobutene isomers in about 80% yield. In a subsequent stage they are converted to 1,4-dicyano-2-butene by treatment with aqueous hydrocyanic acid in presence of calcium carbonate and cuprous compounds [3.156]:

$ClCH_2-CH=CH-CH_2Cl$
1,4-Dichloro-2-butene

$CH_2=CH-CHCl-CH_2Cl$
3,4-Dichloro-1-butene

$\xrightarrow[-2HCl]{+2HCN}$ $NC-CH_2-CH=CH-CH_2-CN$

The 1,4-dicyano-2-butene can be hydrogenated on noble metal catalysts to adiponitrile and, in a further reaction stage, catalytically hydrogenated to HMDA [3.156]. Dilute hydrochloric acid is a by-product of this process. After neutralization with NaOH it contaminates the waste water with a large amount of salt. Du Pont is running two plants in the USA utilizing this process, one in Orange, TX (130,000 t/a HMDA) and one in Belle, WV (90,000 t/a HMDA). Since the start-up of a new plant (see below) these have only rarely been in operation.

Du Pont has developed a chlorine-free synthesis of adiponitrile by catalytical addition of hydrocyanic acid to butadiene in the liquid phase:

$CH_2=CH-CH=CH_2$ $\xrightarrow{+HCN}$ $\xrightarrow{\text{isomers}}$ $CH_2=CH-CH_2-CH_2-CN$ $\xrightarrow{+HCN}$ $NC-(CH_2)_4-CN$

This two-stage catalytic conversion takes place at 100 °C at atmospheric pressure. Nickel complex-catalysts with phosphine or phosphite ligands, and with added metal chlorides are believed to be used for the addition and isomerisation reactions [3.157]. This process makes the chlorine-alkali-electrolysis unnecessary and thus consumes about 25% less energy than the old method. The waste problems are also not so great.

Du Pont began operating the first plant of this type in 1977 in Orange, TX, with a capacity of 120,000 t/a adiponitrile (equivalent to 130,000 t/a HMDA). A year later a joint venture with Rhône Poulenc (Butachimie) resulted in an almost identical plant in Chalampé, France, with a capacity of 110,000 t/a adiponitrile (corresponding to 120,000 t/a HMDA) [3.148]. Up to now Du Pont has shown no interest in awarding additional licenses.

The base chemicals for preparing longer-chain polyamides can also be obtained by cyclic di- and trimerisation of butadiene according to Ziegler chemistry. Two butadiene molecules combine to give 1,5-cyclooctadiene (COD) on triethylaluminum and a nickel compound as co-catalyst. Polyamide 8 can be made from this via cyclo-octan, -octanol, -octanone, -octanone oxime and capryllactam. The trimerisation to 1,5,9-cyclododecatriene (CDT) is more important, this being the starting material for polyamide 12. Therefore ethylaluminum sesquichloride and a titanium compound as co-catalyst are used [3.158].

Butadiene — Trimerisation according to Wilke / Ziegler → Cyclododecatriene — Hydrogenation →

Cyclododecane — Oxidation → Cyclododecanol — Dehydrogenation →

Cyclododecanone — Oximation → Cyclododecanonoxime — Beckmann rearrangement → Laurolactam

— Polymerisation → $\{CO-NH-(CH_2)_{11}\}$

Polyamide-12

Hüls, among others, developed polyamide 12 from the middle of the 1950s and built a large scale plant in Marl. This has been extended step by step to today's 14,000 t/a capacity. The plant is also suitable for the production of polyamide 6,12. The capacity for laurolactam is said to be 20,000 t/a.

COD and CDT have also been produced in Japan since the middle of the 1960s. This is at Toyo Soda, utilizing their own process [3.159]. Shell also produces these intermediates in the USA and France.

Dodecanedioic acid can be made via CDT and is used by Hüls to produce polyamide 6,12 by reaction with HMDA.

Du Pont also produced, up to 1982, a special nylon fibre on the basis of CDT (Quiana, an artificial silk for ties).

3.4.4 1,4-Butanediol and Derivatives

1,4-Butanediol can be produced from butadiene in three stages using processes of Mitsubishi [3.160] and of BASF [3.161]:

$$CH_2=CH-CH=CH_2 \quad + \quad 2\ CH_3-COOH \xrightarrow{+O_2} CH_3-COOCH_2-CH=CH-CH_2-COOCH_3 \ + \ H_2O$$

$$CH_2OH-CH_2-CH_2-CH_2OH \ + \ 2\ CH_3-COOH \xrightarrow{+H_2O}$$

$$\downarrow{+H_2}$$

$$CH_3-COOCH_2-CH_2-CH_2-CH_2-COOCH_3$$

$$-H_2O \downarrow$$

$$\begin{array}{c} CH_2-CH_2 \\ | \qquad\quad O \\ CH_2-CH_2 \end{array} \quad + \quad 2\ CH_3-COOH \xrightarrow{+H_2O}$$

In the first stage butadiene is converted in the liquid phase to 1,4-diacetoxy-2-butene with acetic acid and oxygen at 80 °C on Pd-Te-catalysts. This is hydrogenated on a Ni-Zn-catalyst at 80 °C and 60 bar. In the final step the saturated diacetate is hydrolysed in the presence of sulfuric acid. Under mild conditions this yields preferentially tetrahydrofuran (THF) but under more severe conditions, 1,4-butanediol.

1,4-Butanediol can also be cyclised to THF in the presence of steam. The reaction conditions of all the steps mentioned are so advantageous that the total yield of 1,4-butanediol/THF reaches 90%. Since 1982, Mitsubishi Chemical Industries in Japan is the only company in the world operating a 1,4-butanediol plant based on butadiene (15,000 t/a).

Since 1971, Toyo Soda in Shin-Nanyo, Japan have been producing 1,4-diacetoxy-2-butene by reaction of 1,4-dichloro-2-butene with sodium acetate. 1,4-Dichloro-2-butene is a co-product of the chloroprene preparation derived from butadiene (see Sect. 3.4.1.2). It may then be converted into THF and 1,4-butanediol [3.162].

THF is produced by catalytic hydrogenation of MA (at least 99.9% pure) by Nippon Hydrofuran (a joint venture of Mitsubishi Chemical Industries and Mitsubishi Petrochemical) and Hokkaido Organic (joint venture of Dai Nippon Ink and Hokkaido Soda). They obtain the MA from butadiene-containing raffinate II [3.104]. The yield of THF, based on MA, is believed to be 85 to 95%. The principal by-product is γ-butyrolactone, plus a little 1,4-butanediol, which can be utilized readily. The yield of 1,4-butanediol can be appreciably increased by hydrolysing THF in the presence of sulfuric acid or an acidic ion exchanger. Secondary products from γ-butyrolactone are polyamide 4 (via butyrolactam), N-methylpyrrolidone (NMP), and 2-pyrrolidone. These have favourable growth prospects [3.163].

Reactions of butadiene with peroxides to give diperoxybutenes or dihydrofuran have been described in recent patents of Shell and Chevron respectively; these products can be converted by hydrogenation into 1,4-butanediol or THF [3.156].

At the end of 1986 the total world capacity for 1,4-butanediol was about 425,000 t/a. The share of the FRG was 110,000 t/a from BASF and 70,000 t/a from GHC (GAF-Hüls Corporation). USA producers are Du Pont (80,000 t/a), GAF (55,000 t/a), and BASF-Wyandotte (70,000 t/a) [3.164]. These use the Reppe process via acetylene and formaldehyde exclusively (see Sect. 2.2.4.2).

1,4-Butanediol is a versatile intermediate for preparing polyurethanes and polyesters. Linear or cross-linked polyurethanes are formed by polyaddition of various isocyanates. The linear polybutylene terephthalate (PBTP) is obtained by polycondensation with dimethyl terephthalate (terephthalic acid esterified with methanol) or with oxidatively esterified p-xylene [3.165].

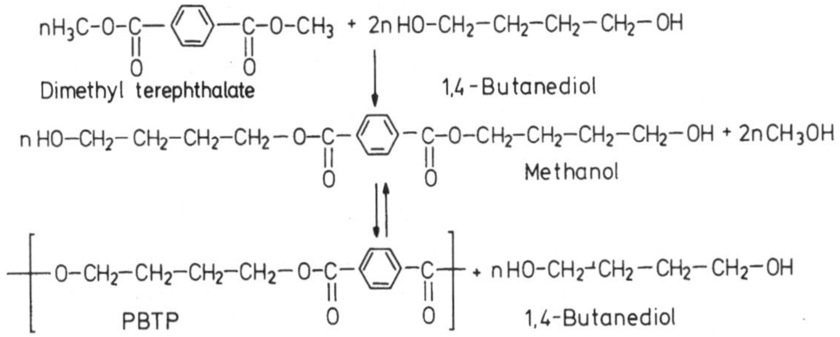

$$n H_3C-O-\underset{O}{\overset{O}{C}}-\langle\text{benzene}\rangle-\underset{O}{\overset{O}{C}}-O-CH_3 + 2n\, HO-CH_2-CH_2-CH_2-CH_2-OH$$

Dimethyl terephthalate   1,4-Butanediol

$$n\, HO-CH_2-CH_2-CH_2-CH_2-O-\underset{O}{\overset{O}{C}}-\langle\text{benzene}\rangle-\underset{O}{\overset{O}{C}}-O-CH_2-CH_2-CH_2-CH_2-OH + 2n\, CH_3OH$$

Methanol

$$\left[-O-CH_2-CH_2-CH_2-CH_2-O-\underset{O}{\overset{O}{C}}-\langle\text{benzene}\rangle-\overset{O}{C}-\right]_n + n\, HO-CH_2-CH_2-CH_2-CH_2-OH$$

PBTP   1,4-Butanediol

Derived specialities include: thermoplastic, elastomeric polyether-ester-block-copolymers with soft ether segments from condensed butanediols and ester blocks with long-chain aliphatic dicarboxylic acids (e. g. Arnitel of Arco and Hytrel from Du Pont, with sebacic acid as a butadiene speciality); and PBTP alloys with polycarbonates (e. g. Makroblend from Bayer or Xenoy XD from General Electric, with bisphenol A).

Producers of polyurethanes in the FRG are: Elastogran, Lemförde, belonging to the BASF group; and Bayer, Leverkusen. Producers of PBTP are: BASF (Ultradur B); Bayer (Pocan B); Hüls Troisdorf (Dynalit); and Hüls (Vestodur B).

3.4.5 Anthraquinone

Bayer have developed a three-stage process for anthraquinone, based on naphthalene and butadiene [3.163]:

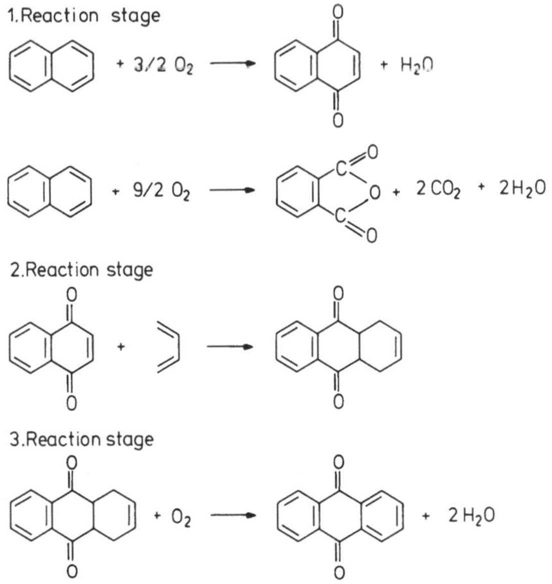

1. Reaction stage

$$\text{naphthalene} + 3/2\, O_2 \longrightarrow \text{naphthoquinone} + H_2O$$

$$\text{naphthalene} + 9/2\, O_2 \longrightarrow \text{phthalic anhydride} + 2\, CO_2 + 2\, H_2O$$

2. Reaction stage

$$\text{naphthoquinone} + \text{butadiene} \longrightarrow \text{tetrahydroanthraquinone}$$

3. Reaction stage

$$\text{dihydroanthraquinone} + O_2 \longrightarrow \text{anthraquinone} + 2\, H_2O$$

Naphthalene is oxidised in the first step to a mixture of 1,4-naphthoquinone and phthalic anhydride. The naphthaquinone in the mixture is converted to tetrahydroanthraquinone by Diels-Alder reaction with butadiene. In the final stage this quinone is oxidised to anthraquinone. Phthalic anhydride and unused naphthalene are distilled off, leaving pure anthraquinone as the bottom product.

A few years ago Bayer erected a plant based on this process, with an anthraquinone capacity of 12,000–15,000 t/a, sited in the new works of Schelde Chemie Brunsbüttel. However this plant has never gone into production for technical and economic reasons [3.167].

About 85% of the world production of anthraquinone stems from the oxidation of anthracene. Its world capacity is estimated to be about 30,000 t/a at present. Bayer are the world's largest producer (11,000 t/a) and Ciba-Geigy produce 4000 t/a [3.168].

3.4.6 Sulfolane

Sulfolane is the trivial name for tetrahydrothiophene-1,1-dioxide, ($C_4H_8SO_2$). It is employed by Shell for separating aromatics and paraffins [3.169]. In Shell's own process, it is made by the reaction of butadiene with SO_2. This yields 3-sulfolene which is hydrogenated to sulfolane [3.170].

$$CH_2=CH-CH=CH_2 \ + \ SO_2 \ \longrightarrow \ \underset{O^{\nearrow S}\diagdown O}{\bigcirc} \ \overset{H_2}{\longrightarrow} \ \underset{O^{\nearrow S}\diagdown O}{\bigcirc}$$

Sulfolane decomposes into the starting materials butadiene and SO_2 on heating to about 125 °C. World production capacity is estimated to be over 7000 t/a, of which Shell, Stanlow in Great Britain, is the main supply centre for Europe (4000 t/a).

3.4.7 Other Butadiene Specialities

Butadiene forms heterocyclic compounds with certain elements, analogously to the formation of sulfolane. Thiophene, as a sulfur compound, is the most important of these. It is obtained with 40% yield in the reaction between butadiene and sulfur at 450 °C [3.171]:

$$CH_2=CH-CH=CH_2 \ + \ 2S \ \longrightarrow \ \underset{S}{\bigcirc} \ + \ H_2S$$

Thiophene is the starting material for preparing numerous pharmaceuticals, e. g. the antibiotic Cefalotin (Eli Lilly) and the antihistamine Thenalidin [3.172]. It is also used as a solvent and in the production of thiophene-phenol-formaldehyde resins, dyes and plant protection agents.

Butadiene and MA react in a Diels-Alder addition to give tetrahydrophthalic anhydride (THPA). This is added in small amounts to heat- and fire-resistant unsaturated polyester and alkyd resins [3.171].

Hüls also produces an MA-modified BR-polyoil, suitable for electric dip varnishing [3.154].

A special terpolymer of butadiene, styrene and 2-vinylpyridine has proved useful for improving adhesion of tire cord in automobile and aircraft tyres which are subjected to great stress. The 2-vinylpyridine is mainly the secondary product from 2-methylpyridine (α-picoline), a constituent of hard coal tar [3.173]. Hüls and Bayer are the producers in the FRG.

An important field of application for butadiene is in the preparation of complex catalysts with Ni, Co, Pd or Fe. These catalysts are used in oligomerisation reactions [3.174]. The Ni-complexes of butadiene are particularly suitable for dimerisation of the butadiene to 1,3,6- and 1,3,7-octatrienes. The latter product can be cyclised to xylene and ethyl benzene by dehydrogenation. This is, however, only of theoretical interest on account of the excessively high processing costs. More valuable are the products of hydrogenation, 1,6- and 1,7-octadiene, which can be converted by OXO-synthesis or addition of HCN, with subsequent hydrogenation, to C_{10}-diols or C_{10}-diamines, respectively. These C_{10}-diamines are intermediates for deriving special polyesters or polyamides. Further important specialities are the hydrogenated hydrocarboxylated products of the octadienes mentioned:

$$
2\ CH_2{=}CH{-}CH{=}CH_2 \Big\langle
\begin{array}{l}
\xrightarrow[\ +\,CO\,+\,H_2O\ ]{}\ \xrightarrow[\ +\,2\,H_2\ ]{}\ CH_3(CH_2)_7COOH \quad \text{Pelargonic acid}\\[2ex]
\xrightarrow[\ +\,2CO\,+\,2H_2O\ ]{}\ \xrightarrow[\ +\,2\,H_2\ ]{}\ HOOC{-}(CH_2)_8{-}COOH \quad \text{Sebacic acid}
\end{array}
$$

One of the uses of pelargonic acid is as a starting material for producing heat-resistant lubricants [3.175]. Sebacic acid is suitable for making special polyesters (see Sect. 3.4.4) and is combined with hexamethylenediamine in polyamide 6,10 (Ultramid S of BASF, Technyl of Rhône-Poulenc). The co-product from the preparation of sebacic acid at BASF, a mixture of 2-ethylsuberic and 2,5-diethyladipic acids (so-called isosebacic acid) is processed to plasticizers.

As early as in the 1960s, Toyo Soda proposed building up a speciality chemicals complex based on COD and CDT (see Sect. 3.4.3). In this way the aim of producing larger production amounts was accomplished by broadening the production program with these valuable but expensive intermediates. Consequently, a reduction in the manufacturing and marketing costs was expected [3.159]. Some of these propositions, which have for the most part not been able to make the technical-economic breakthrough, are given in Fig. 3.29. Shell hydrogenate COD to cyclooctene in Berre. This is submitted to metathesis with ethylene to yield 1,5-hexadiene and 1,9-decadiene [3.176].

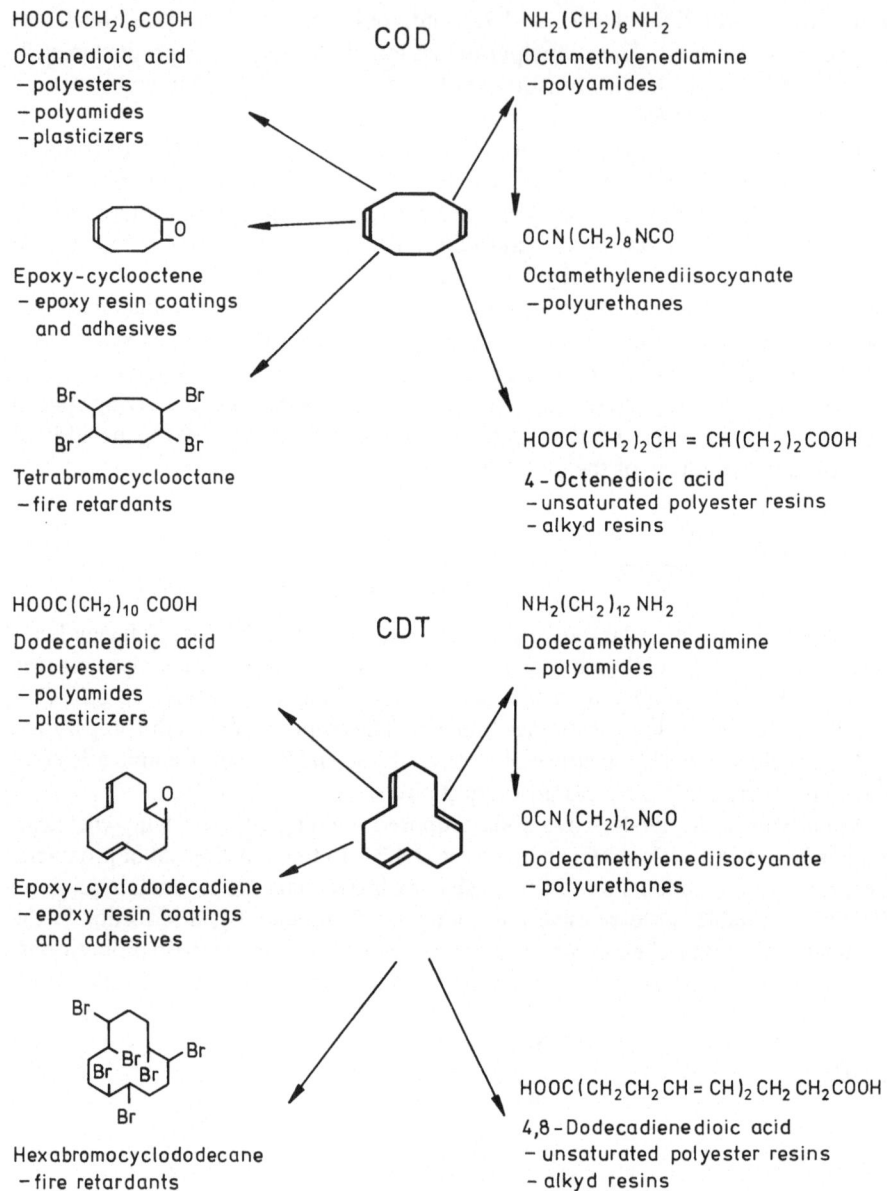

HOOC(CH₂)₆COOH

Octanedioic acid
– polyesters
– polyamides
– plasticizers

Epoxy-cyclooctene
– epoxy resin coatings
  and adhesives

Tetrabromocyclooctane
– fire retardants

COD

NH₂(CH₂)₈NH₂

Octamethylenediamine
– polyamides

OCN(CH₂)₈NCO

Octamethylenediisocyanate
– polyurethanes

HOOC(CH₂)₂CH = CH(CH₂)₂COOH

4 - Octenedioic acid
– unsaturated polyester resins
– alkyd resins

HOOC(CH₂)₁₀ COOH

Dodecanedioic acid
– polyesters
– polyamides
– plasticizers

Epoxy-cyclododecadiene
– epoxy resin coatings
  and adhesives

Hexabromocyclododecane
– fire retardants

CDT

NH₂(CH₂)₁₂ NH₂

Dodecamethylenediamine
– polyamides

OCN(CH₂)₁₂ NCO

Dodecamethylenediisocyanate
– polyurethanes

HOOC(CH₂CH₂CH = CH)₂ CH₂ CH₂COOH

4,8 - Dodecadienedioic acid
– unsaturated polyester resins
– alkyd resins

Fig. 3.29. Innovative derivatives of cyclooctadiene (COD) and cyclododecatriene (CDT)

4 Processing Scheme for C_4-Hydrocarbons

4.1 Method of Ascertaining the Processing Scheme

The processing scheme is intended to give information about the qualitative and quantitative relations between the raw materials, the base chemicals, intermediates, and the end products as last products within the production chains. It is the basis for analyzing and forecasting the supply and consumption developments of the particular base, intermediate, and industrial or speciality chemicals.

As long as there exists a strong demand for a raw material, base chemical or intermediate the processing of which is highly profitable, the competitive ability of the other producers who are dependent on that material can be impaired as a result of supply shortages. This is true, for instance, when the processor with the favorable supply position has only low processing costs and the secondary products are high-priced.

The input-output scheme begins with a list of the present and future possible uses, according to the different base chemicals, as seen for the C_4-hydrocarbons in the table of contents to Chap. 3. One has to develop a product family tree (dendrogram), a production hierarchy displaying the scheme of consumption in the product generations. This provides an insight into the final stages of chemical production and the most important products which can be manufactured from the industrial and speciality chemical end products.

It is also customary to relate the end products of the chemical industry to the markets in which they will finally be used. This quantitative scheme, orientated according to the business line or markets, forms the basis for global forecasts of substitution developments between different products or market segments. This is influenced by changes of structure and the state of the economy of the end markets or industries in which the products will find their final use. Thus these consumption patterns provide comprehensive information on special sectors.

Production data are required for the base, intermediate or end product chemicals in order to quantify the input-output streams. The source of information for this is primarily the secondary statistics issued by government departments or industrial associations (see Table 4.1). A disadvantage of the secondary statistics is that many products are missing or are summarized under headings where components are not generally known. The production processes of the various products are usually not revealed in these statistics, unless the relevant process can be deduced from the special nature of the product. Authorities active in the field of Chemical Market Research have dealt with the special problems of interpreting secondary statistical data [4.1].

Table 4.1. Important chemical production and trade statistics

Name	Content	Region covered	Sequence	Institute	Basis of information
Statistisches Bundesamt (Fachserie 4, Reihe 3.1)	Production amounts and values, foreign trade data	FRG plus some international data	monthly quarterly yearly	Statistisches Bundesamt, Wiesbaden	Reports of the producers sent to the statistical authorities
MW-Statistik	Capacities, production and consumption of refinery-raw materials and products	FRG plus some international data	yearly	Mineralölwirtschaftsverband, Hamburg	Reports of refinery managers
VCI-Statistik, internal*, external	Capacities, production and consumption of chemical products	FRG	yearly	Verband der Chemischen Industrie, Frankfurt	Government statistics, questionnaires answered by the chemical companies
The Chemical Industry	Production and foreign trade of chemical products	European Community countries	yearly	OECD Publications office, Paris	Reports of national associations
CEFIC-statistics	Capacities, production and consumption of petrochemical products	W. Europe	yearly	CEFIC, Brussels	Reports of national associations
BIT-data*	Capacities, production and foreign trade of chemical products	W. Europe	half-yearly	Bureau Int. Technique, Brussels	Reports of national associations
IISRP-statistics	Capacities, production and use of synthetic rubber	worldwide	yearly	International Institute of Synthetic Rubber Producers, Houston	Studies of synthetic rubber producers and processors
VKE-Zahlen*	Classification of inland plastic sales according to principal fields of application	FRG	yearly	Verband der Kunststofferzeugenden Industrie, Frankfurt	Reports of plastic producers via a trust institution

* only for members of the relevant organisations

Table 4.2. Studies of chemical technology and chemical markets (all freely available)

Title	Content	Region covered	Sequence	Institute	Basis of information
Chemical Economics Handbook (CEH)/ World Hydrocarbons	Capacities, production, foreign trade, prices and use; Process engineering; cost data of chemical products	worldwide, USA mainly	serial deliveries	Stanford Research Institute, Menlo Park, CA, USA	Expert research work; statistics of the US tariff commission
Process evaluation/research planning (PERP)	Supply- and demand-amounts, chemical product family trees, process engineering; cost data	USA, Japan, W. Europe	every 2−3 years, revised reports	Chem. Systems, New York, NY, USA	Expert research
Trichem Olefin Price Service	Capacities, production, foreign trade and use of petrochemicals; market forecasts	worldwide	yearly	Trichem Consultants, London	Expert research
CHEMEUROP	Capacities, production, foreign trade and use, classified according to countries and 130 chemical products	Europe	serial deliveries	SEMA, Paris	Evaluation of statistics and literature, questioning by chemical market researchers
Chemical Product Data	Capacities, production, foreign trade and use of all chemical products with information about producers	worldwide	serial deliveries	Chemical Data Services, London	Mainly evaluation of literature information
De Witt Annuals/ Surveys	Capacities, production, foreign trade and use of petrochemicals	worldwide	yearly	De Witt and Co., Houston, TX, USA	Expert research
CMAI-Report and World Analysis	Capacities, production, foreign trade and use of petrochemicals, market forecasts, process costs	worldwide	yearly	Chemical Market Associates, Houston, TX, USA	Expert research
Aromatics and Olefins Report	Capacities, production, use of petrochemicals, market forecasts	W. Europe	yearly	Parpinelli, Tecnon, Milan, Italy	Expert research

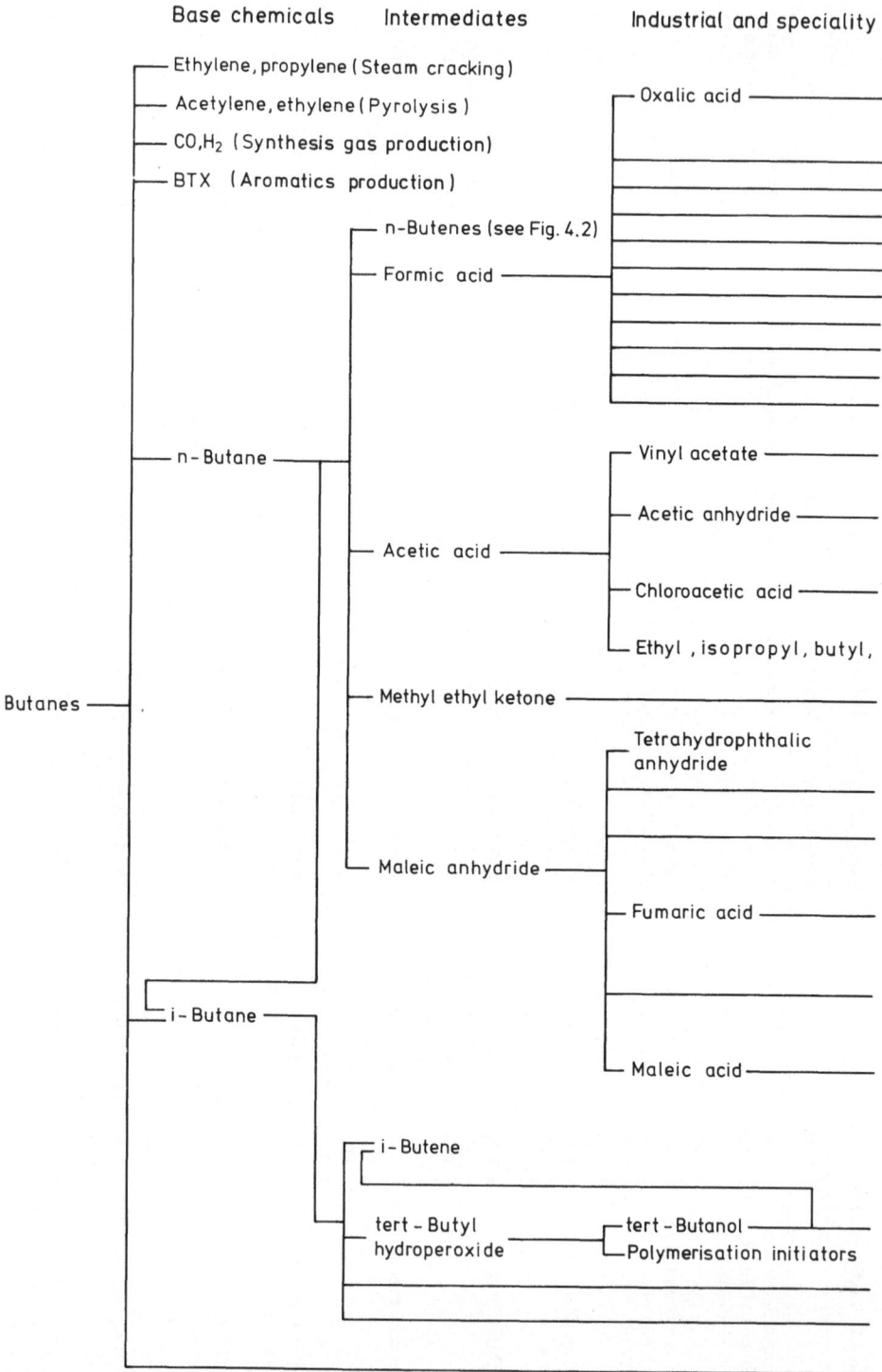

Fig. 4.1. Butane derivatives processing scheme

chemicals   Products

K-,Fe-,NH$_4$-salts of oxalic acid
- Paints and colors (for calico printing, blueprints etc.)
- Cleansers (ink cleansers etc.)
- Leather tanning agents

Aluminium formiate
- Textiles impregnating agents
- Textiles dyestuffs and finishes
- Leather treatment chemicals
- Latex coagulating agents
- Food and animal feed preservation agents
- Pesticides
- Solvents for lacquers and perfumes
- Galvanizing agents
- Auxiliaries for mirror glass production
- Ore flotation agents

VA-ethylene-copolymer — Hot-sealable adhesives
Polyvinylacetate — Dispersed dyestuffs

Cellulose acetate — Cellulose fibers and films
Acetylsalicylic acid — Pharmaceuticals (Aspirin, Phenacetin)

Herbicides 2,4-D and 2,4,5-T — Pesticides
Synthetic coffein — Cola beverages, pharmaceuticals
Thioglycolic acid — Hair waving preparations

ammonium acetate — Solvents for varnishes and adhesives

- Solvents for vinyl and acrylic coating resins, inks
- Solvents for lubricating oils dewaxing

Copolymers
- Cleansers, floor polishes
- Hair sprays, adhesives, pharmaceuticals

BR polyoils — Electrical dipping varnishes

Alkyd resins
- Adhesives, linoleum bonding agents
- Coatings, cable sheathings, varnishes

unsaturated polyester resins
- Profile material, sheets, molded parts
- Filling material

- Lube oil additives, pesticides, plant growth regulators

- Food preservation agents

Tetrahydrofuran/1,4-butanediol
γ-Butyrolactone

Butyrolactam — Nylon-4 — Nylon fibers and plastics
N-Methylpyrrolidone — Extracting agents for dienes
Polyvinylpyrrolidone — Glue, blood substitute solutions (Periston)

GTBA with methanol — Gasoline component
for LDPE, PVC, PS, ABS — Plastic products

Alkylate as gasoline component — Gasoline component

- Propellant, solvent

- Gasoline component
- Heating gas (industry, camping)

More detailed information is furnished by chemical technology and market studies, prepared by research institutes. They are reports issued specially for individuals or groups of companies (see Table 4.2). Information is also collected by the interested companies themselves. Chemical market research congresses and working groups are a fruitful source of information, however, almost exclusively for insiders.

The requirement of base chemicals can be deduced from the production amounts of intermediate and performance chemicals via the specific consumption data related to the particular process. The process yield-data in mole % or wt. % needed for such computations can be taken from the Chemical Technology literature, Chemical Market Research literature, from process licensors or from one's own experience, and can be calculated from the following equation:

$$\text{consumption coefficient} = \frac{\text{molecular weight of the raw material}}{\text{mole yield} \times \text{molecular weight of product}} = \frac{1}{\text{wt. \% yield}}$$

$$\text{yield} = \text{conversion} \times \text{selectivity}$$

The data are generally classified in tables according to secondary products and regions, or, for particular regions, only according to secondary products in a pie chart.

Figures 4.1, 4.2, and 4.4 give the progressive spread of the consumption patterns or utilization schemes for butanes, butenes and butadiene. The supply patterns for the production of butenes and butadiene corresponding to this are given in Fig. 2.1 and in Tables 2.1 and 2.11.

4.2   Analysis of the Processing Scheme for Butane Secondary Products

4.2.1   Butane Derivatives Processing Scheme

Figure 4.1 shows the possible and actually executed processing routes starting from butane, as described in Sect. 3.2, at the present state of technical progress.

4.2.2   Production Amounts of Butane Secondary Products

Statistical data about the amounts of derivative products from butane and of other uses of butane (heating gas) are rarely published. Some sources provide quite extensive information for the year 1984. These figures have been collected and summarised in Table 4.3.

No reliable information is available about the use of butane for heating. The problem here is that liquefied gas for heating purposes, as quoted in the statistics, can contain a widely varying and not precisely given content of propane. The amounts used in the gasoline sector, as alkylate gasoline or directly employed as a gasoline component, predominate. Other butane secondary products in 1984 were: TBA (only Arco); butenes (n-butenes from dehydrogenation of n-butane and i-butenes via Arco TBA or, in Eastern Europe, via dehydrogenation of i-butane); ethylene (steam

Table 4.3. Amounts of butane derivatives produced in the world during 1984 (in 1000 t/a) [4.2]

Derivative	North America	South America	Western Europe	Eastern Europe	Rest of world	Total
Ethylene	600	–	200	–	n. a.	> 800
Acetylene	–	–	30	n. a.	–	> 30
CO/H$_2$	n. a.	n. a.	n. a.	n. a.	n. a.	n. a.
n-Butenes	30	60	–	600	40	730
i-Butene	250**	–	50*	50	–	350
Acetic acid**	200	–	–	–	–	200
Maleic anhydride	170	–	–	n. a.	n. a.	> 170
TBA***	530	–	410	–	–	940
Butene alkylate gasoline	26360	580	2000	620	1090	30650
Propene alkylate gasoline	7340	490	1700	520	920	10970
Directly blended with gasoline	12000	n. a.	n. a.	n. a.	n. a.	60000
Heating gas	1500	n. a.	n. a.	n. a.	n. a.	20000

 * only through dehydration of Arco-TBA, with 53% by wt. propene oxide as co-product, based on
  TBA
 ** only Celanese, Pampa, TX, with 16% by wt. (based on the acetic acid) methyl ethyl ketone, 9%
  formic acid, 7% ethyl acetate and 1% ethanol
 *** only Arco-GTBA for the motor fuel sector, with 40% by wt. propene oxide as co-product, based
  on TBA

cracking); maleic anhydride (only in the USA on butane basis, from Amoco, Ashland, Denka and Monsanto); and acetylene (only in the FRG at Hüls, utilization of residues from raffinates from preparation of MTBE and TBA in the electric arc).

4.2.3 Specific Butane Consumption Coefficients for the Secondary Products

Butane consumption coefficients for the most important secondary products and processes are given in Table 4.4.

4.2.4 Processing Scheme for Butanes

The world consumption of butanes, classified according to secondary products and regions, is given in Table 4.5. It results from combining the information in Tables 4.3 and 4.4.

4.3 Analysis of the Processing Scheme for Butene Secondary Products

4.3.1 Butene Derivatives Processing Scheme

The technically developed routes to butene secondary products, described in Sect. 3.3, are shown in Fig. 4.2.

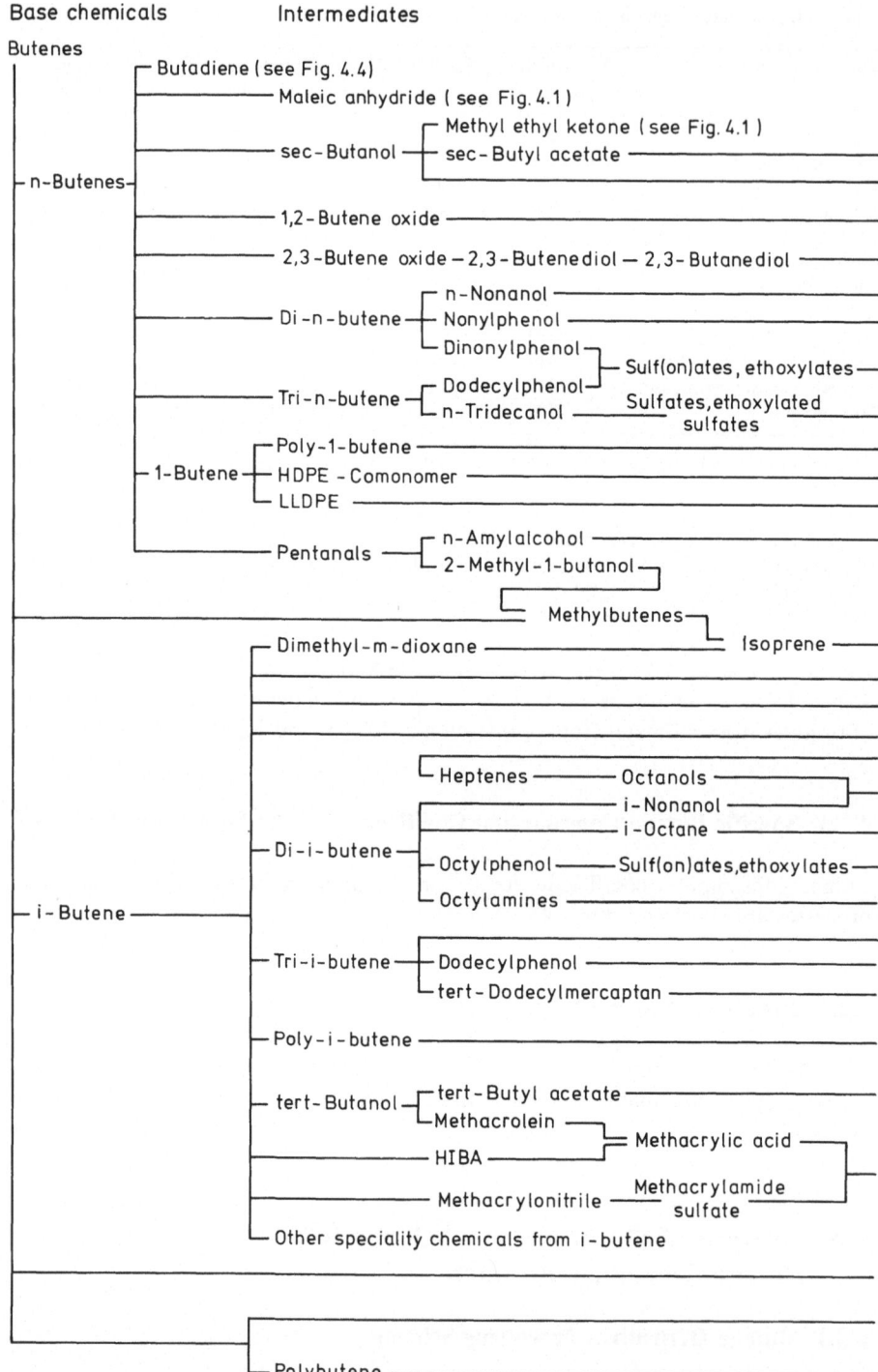

Fig. 4.2. Butene derivatives processing scheme

Industrial and speciality   Products
        chemicals

————————————————————— Solvent for varnishes and adhesives
————————————————————— Solvent for nitrocellulose (varnishes, celluloid)
___ Solvent for chlorinated_____ Solvent (metal degreasing etc.)
      hydrocarbons, stabilizer
—— Elastomer molding resins ——————— Tires, rollers, couplings, bumpers
—— Phthalates as plasticizers ——————— Flexible PVC plastic products
—— Alkylphenolic resins ————————— Varnishes, adhesives, laminates

—— Surfactants ——————————— Detergents

—— Surfactants ——————————— Detergents
————————————————————— Plastic products (floor heating pipes etc.)
————————————————————— HDPE film products (bags etc.)
————————————————————— LLDPE film products
————————————————————— Solvents

┌— Polyisoprene ————————————— Synthetic rubber for tires
│   ┌— Inner tubes of tires
├— Butyl rubber ———————————┤
│   └— Inside material of tubeless tires
—— MTBE ——————————————— Gasoline component, solvent
—— Polymer gasoline ————————┐——— Gasoline component
___ Phthalate / adipic acid _____ Flexible PVC plastic products
          plasticizers
————————————————————— Gasoline component
—— Surfactants ——————————— Detergents
—— Rubber chemicals ——————— Synthetic elastomers
————————————————————— Lube oil additive
—— Alkylphenolic resins ——————— Bonding agents, laminates, molded parts
—— NBR, SBR and latexes ——————— Synthetic elastomer products, textile coatings
                                    ┌— Viscosity-index improvers
————————————————————┤
                                    └— Adhesives, sealing tapes
————————————————————— Gasoline additive

                  ┌— MBS ——————————— High impact PVC plastic parts and profiles
—— MMA ———┤
                  └— PMMA ——————————— Polymethyl methacrylate plastic products

                                    ┌— Gasoline component
————————————————————┤
                                    └— Heating gas (refinery)
___ Alkylate as gasoline _____ Gasoline component
          component
                                    ┌— Additive for lubricants and lube oil
————————————————————┤— Adhesives, sealing components
                                    └— Film sealings, cable sheathing

Table 4.4. Butane consumption coefficients, t raw material/t product

Product	Process	Coefficient
	Raw Material *n*-Butane	
i-Butane	BP isomerisation	} 1.00
	UOP Butamer	
Ethylene	steam cracker	ca. 3.5
Acetylene	BASF-acetylene	ca. 4.5
	Hüls electric arc	ca. 2.5
CO/H$_2$ (86:14 by wt.)	ICI catalytic reforming	ca. 0.5
n-Butenes	Houdry Catofin	} 1.27
	Phillips Star	
	UOP Oleflex	
Butadiene	Houdry Catadiene	1.69
	Phillips Butadiene	1.43
Formic acid	}	} n. a.
Acetic acid	Celanese air oxidation	
Methyl ethyl ketone	direct oxidation	
	(Celanese, Gulf Oil, Hüls)	n. a.
Maleic anhydride	Halcon/SD	1.17
	Alusuisse Alma	0.95
Fumaric acid	Alusuisse Alma	0.83
BTX-aromatics	BP/UOP Cyclar	0.86
	Raw material *i*-Butane	
i-Butene	Houdry Catofin	} 1.25
	Phillips Star	
	UOP Oleflex	
	Arco propene oxide	1.26
tert-Butyl-hydroperoxide	Arco propene oxide	0.74
tert-Butanol	Arco propene oxide	0.90
Butene alkylate	Kellogg-, Stratco-H$_2$SO$_4$	0.60
gasoline	UOP-, Phillips-HF	0.63
Propene alkylate	Kellogg-, Stratco-H$_2$SO$_4$	0.60
gasoline	UOP-, Phillips-HF	0.70
Methyl methacrylate	Halcon dehydrogenation with oxidation	0.92

4.3.2 Production Amounts of Butene Secondary Products

Table 4.6 shows that the principal butene secondary products are the gasoline components alkylate, polymer gasoline, and MTBE. Under TBA are given only production amounts for its processing to *tert*-butyl acetate and methacrolein, not, however, for processes in which TBA is an intermediate in the production of pure *i*-butene for producing other specialities. The data in Table 4.6 have been calculated from Table 4.8 by means of Table 4.7.

Table 4.5. World processing scheme or end use pattern for butane, 1984 (in 1000 t/a)

Derivative	N. America		S. America		W. Europe		E. Europe		Rest of world		Total	
	n-butane	i-butane	n-butane	i-butane	n-butane	i-butane	n-butane	i-butane	n-butane	i-butane	n-butane	i-butane
Ethylene	2100	–	n.a.	–	700	–	–	–	n.a.	–	>2800	–
Acetylene	–	–	–	–	75	–	n.a.	–	–	–	>75	–
CO/H$_2$	n.a.	–	n.a.	–	n.a.	–	n.a.	–	n.a.	–	n.a.	–
n-Butenes	40	–	75	–	–	–	760	–	50	–	925	–
i-Butene	–	315	–	–	–	65	–	65	–	–	–	445
Acetic acid	n.a.	–	–	–	–	–	–	–	–	–	n.a.	–
Maleic anhydride	200	–	–	–	–	–	n.a.	–	?	–	>200	–
TBA	–	475	–	–	–	365	–	–	–	–	–	840
Butene alkylate gasoline	–	16345	–	360	–	1240	–	385	–	675	–	19005
Propene alkylate gasoline	–	4770	–	320	–	1105	–	340	–	600	–	7135
Directly blended with gasoline } Heating gas	see Table 4.3											

Table 4.6. Amounts of butene derivatives produced in the world during 1984 (in 1000 t/a)

Derivate	North America	South America	Western Europe	Eastern Europe	Rest of world	Total
Butene alkylate gasoline	26360	580	2000	620	1090	30650
Butadiene	45	40	–	380	25	490
MA	–	–	10	–	25	35
SBA	280	15	230	–	110	635
DNB	–	–	15	–	20	35
Isoprene	–	–	–	240	20	260
Butyl rubber	240	–	180	55	45	520
MTBE	750	30	570	135	15	1500
Polymer gasoline	975	75	450	–	225	1725
DIB/TIB	20	–	85	–	15	120
Polybutene	225	5	120	–	25	375
TBA	–	–	15	–	60	75

Table 4.7. Butene consumption coefficients t raw material/t product

Product	Process	Coefficient
	Raw material n-butenes	
Butene alkylate gasoline	Stratco-H₂SO₄ or HF	0.29*
Polymer gasoline	UOP Polygas	0.26
Amyl alcohol	Oxo-synthesis and hydrogenation	0.47
Butadiene	Petro-Tex Oxo-D	1.59
	Phillips O-X-D	1.34
Maleic anhydride	Bayer MA	0.83
SBA	Texaco direct hydration	0.76
	Sulfuric acid hydration	0.90
Methyl ethyl ketone	Texaco SBA oxidation	0.90
	sulfuric acid SBA oxidation	1.06
1,2-/2,3-Butene oxide (15/85 wt. %)	Chlorohydrin	n. a.
DNB/TNB (90/10 wt. %)	Hüls Octol	1.25
	IFP Dimersol X	1.54
LLDPE	UCC Unipol, BP-LLDPE	0.10
HDPE-Comonomer	Hoechst/Uhde HDPE	0.02
Poly-1-butene (atactic and isotactic)	Hüls Polybutene-1	1.00
	Raw material i-butene	
Butene alkylate gasoline	Stratco-H₂SO₄ or HF	0.16*
Polymer gasoline	UOP Polygas	0.13
	IFP Selectopol	1.02
MTBE	Hüls MTBE	0.64
	Neochem MTBE	0.71
TBA (91 wt. %)	Hüls TBA	0.69
	Sulfuric acid hydration	1.07
Polybutene	Cosden Oil PIB	1.10
Poly-i-butene	BASF PIB	1.15
Butyl rubber	Exxon IIR	1.05
DIB/TIB (80/20 wt. %)	Bayer oligomerisation	1.40
Isoprene	Isoprene (IFP, Bayer, Kuraray)	1.03
Methyl methacrylate	Mitsubishi Rayon (via methacrolein)	0.88
	Escambia (via HIBA)	0.99
	Asahi (via methacrylonitrile)	n. a.

* without prior MTBE process

4.3.3 Specific Butene Consumption Coefficients for the Secondary Products

Table 4.7 lists the butene consumption coefficients for the most important processes. Evaluation of uniform consumption coefficients for alkylate and polymer gasolines is made difficult by the raw material flexibility of the process. Where refineries have already largely gone over to using MTBE plants for preparing alkylate gasoline, modified consumption coefficients of 0.14 for i-butene and 0.31 for the n-butenes must be used. This applies especially to N. America and W. Europe but not for the German Federal Republic alone. The assumptions given in Sect. 3.3.1.1 apply to polymer gasoline as far as the composition of the raw materials and the degree of conversion are concerned.

The processes for poly-i-butene, butyl rubber and isoprene in particular depend on the use of pure i-butene. Its preparation from fractions containing i-butene is unavoidably accompanied by loss of material. When i-butene which has been obtained by dehydration of Arco-TBA is used, these losses cannot be allocated because the raw material basis is i-butane, or they could be taken into account by adding 10 to 15% when TBA comes from direct hydration and 20 to 25% when hydration is performed using sulfuric acid. This differentiation and extra costs are, however, not taken into consideration here.

4.3.4 Processing Scheme for Butenes

Table 4.8 gives the world consumption of i- and n-butenes classified according to secondary products and regions. About 50% of the world supply of butenes are used to produce alkylate gasoline, and about one third is employed for heating or in direct blends with gasoline.

When considering only the FRG (Table 4.9 and Fig. 4.3) it is evident that an above-average amount of n-butenes (68%) is either burnt or directly mixed with gasoline. This fits in with a relatively small production of alkylate gasoline.

4.4 Analysis of the Processing Scheme for Butadiene Secondary Products

4.4.1 Butadiene Derivatives Processing Scheme

The technically established routes to butadiene secondary products are shown in Fig. 4.4. These products were discussed in Sect. 3.4.

4.4.2 Production Amounts of Butadiene Secondary Products

Table 4.10 contains a list of the world production of butadiene secondary products where these are 300,000 t/a or more. All the important artificial rubbers and latices, and also ABS/MBS and HMDA belong to these. EPDM could not be quoted because

Table 4.8. World processing scheme or end use pattern for butenes, 1984 (in 1000 t/a) [4.3]

Derivative	N. America		S. America		W. Europe		E. Europe		Rest of world		Total	
	i-butene	n-butenes	i-butene	n-butenes	i-butene	n-butenes	i-butene	n-butenes	i-butene	n-butenes	i-butene	n-butenes
Alkylate gasoline	3690	8170	90	170	280	620	100	180	175	315	4335	9455
Polymer gasoline	120	270	10	20	60	120	–	–	30	60	220	470
Heating gas, blends with gasoline	410	810	665	1435	380	1330	190	720	1115	2035	2760	6330
MTBE	500	–	20	–	370	–	90	–	10	–	990	–
TBA	–	–	–	–	10	–	–	–	40	–	50	–
Di-, tri-i-butene	30	–	–	–	120	–	–	–	20	–	170	–
Butyl rubber	250	–	–	–	190	–	60	–	50	–	550	–
Polybutenes	250	–	5	–	130	–	–	–	30	–	415	–
Isoprene	–	–	–	–	–	–	250	–	20	–	270	–
1,3-Butadiene	–	60	–	60	–	–	–	600	–	40	–	760
1-Butene in polyolefines	–	180	–	10	–	50	–	–	–	40	–	280
SBA	–	250	–	15	–	200	–	–	–	100	–	565
Miscellaneous	100	110	10	20	100	80	10	–	10	30	230	240
Total	5350	9850	800	1730	1640	2400	700	1500	1500	2620	9990	18100

Table 4.9. Processing scheme or end use pattern for butenes in the FRG, 1985 in 1000 t/a

End product	Firm	*n*-butenes	End product	Firm	*i*-butene
Alkylate gasoline	OMW	87	Alkylate gasoline	OMW	48
Polymer gasoline	Shell	n. a.	Polymer gasoline	Shell	n. a.
	Texaco	6		Texaco	3
	URBK	20		URBK	10
1-Butene for HDPE/ LLDPE	Hüls	25*	MTBE	Hüls	96
				Texaco	10
SBA	Texaco	46	TBA (91 wt. %)	Hüls	34**
n-Octenes	BASF	15	DIB/TIB	EC	80
	Hüls	5	PIB	BASF	11
MA	Bayer	12		Esso	40
Heating gas, blends				ROW	9
with gasoline		440	Heating gas, blends		
			with gasoline		203
Total		656	Total		656

* about 60% of this for export to other European countries, Argentina, and the USA
** about 24,000 t/a of this for producing speciality end products

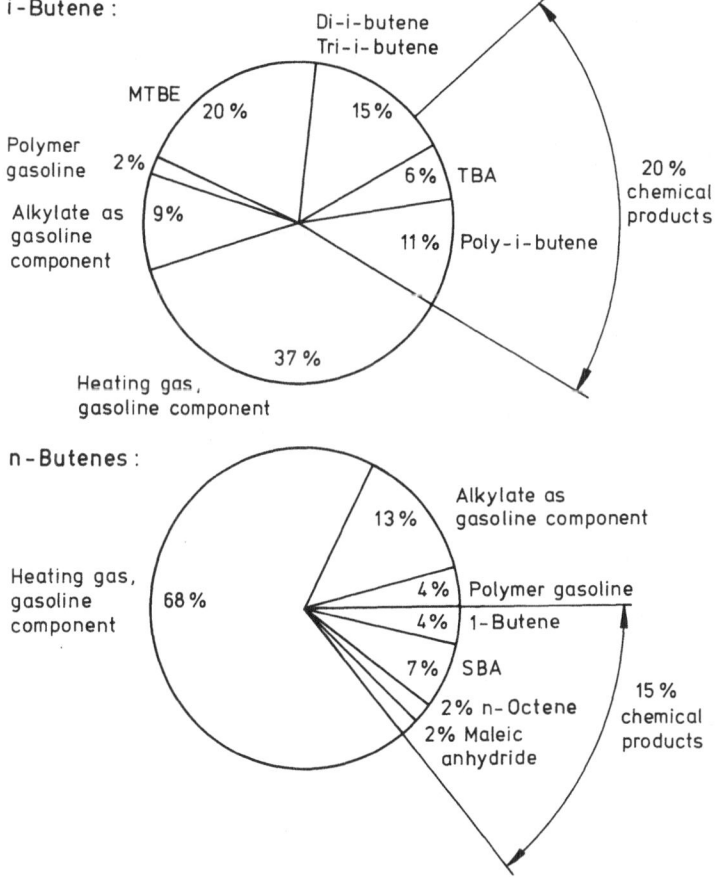

Fig. 4.3. Consumption patterns for *n*-butenes and *i*-butene in the FRG 1985 (in wt. %)

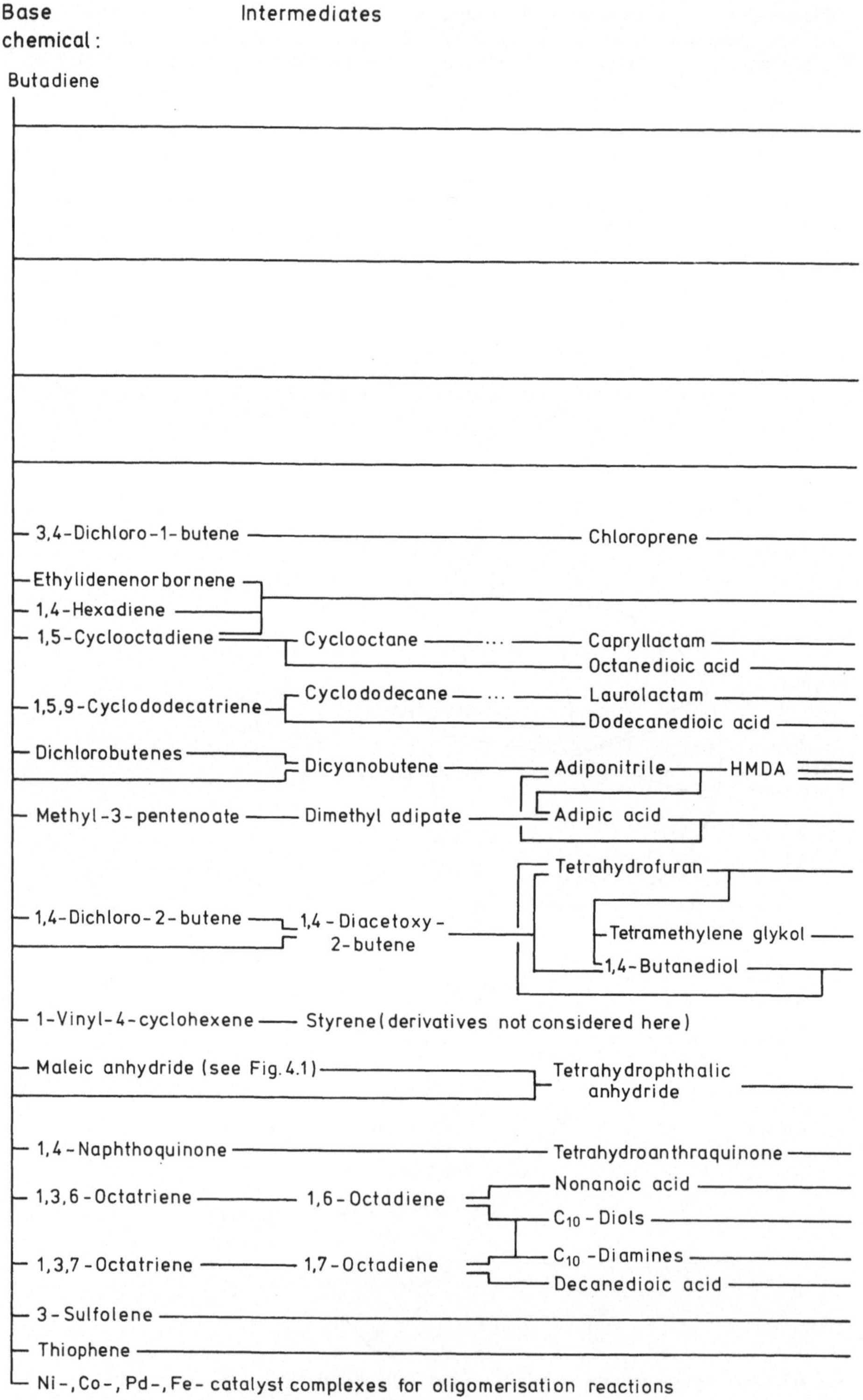

Base   Intermediates
chemical :

Butadiene

3,4-Dichloro-1-butene —————————————— Chloroprene —————————

Ethylidenenorbornene
1,4-Hexadiene
1,5-Cyclooctadiene ═══ Cyclooctane ——— ... ——— Capryllactam ————
                                              Octanedioic acid ————
                        Cyclododecane —— ... —— Laurolactam —————
1,5,9-Cyclododecatriene   Dodecanedioic acid ———

Dichlorobutenes ——— Dicyanobutene ———— Adiponitrile ——— HMDA ═══

Methyl-3-pentenoate ——— Dimethyl adipate ——— Adipic acid ——

                                              Tetrahydrofuran —————
1,4-Dichloro-2-butene —— 1,4-Diacetoxy-
                         2-butene ——————— Tetramethylene glykol ——
                                           1,4-Butanediol ————

1-Vinyl-4-cyclohexene ——— Styrene (derivatives not considered here)

Maleic anhydride (see Fig. 4.1) ————————— Tetrahydrophthalic
                                            anhydride —————

1,4-Naphthoquinone ——————————— Tetrahydroanthraquinone ———
                                              Nonanoic acid ————
1,3,6-Octatriene ——— 1,6-Octadiene
                                           C₁₀-Diols ————
                                           C₁₀-Diamines ————
1,3,7-Octatriene ——— 1,7-Octadiene
                                           Decanedioic acid ————
3-Sulfolene ——————————————————————

Thiophene ————————————————————————

Ni-, Co-, Pd-, Fe- catalyst complexes for oligomerisation reactions

Fig. 4.4. Butadiene derivatives processing scheme

Industrial and speciality chemicals   Products

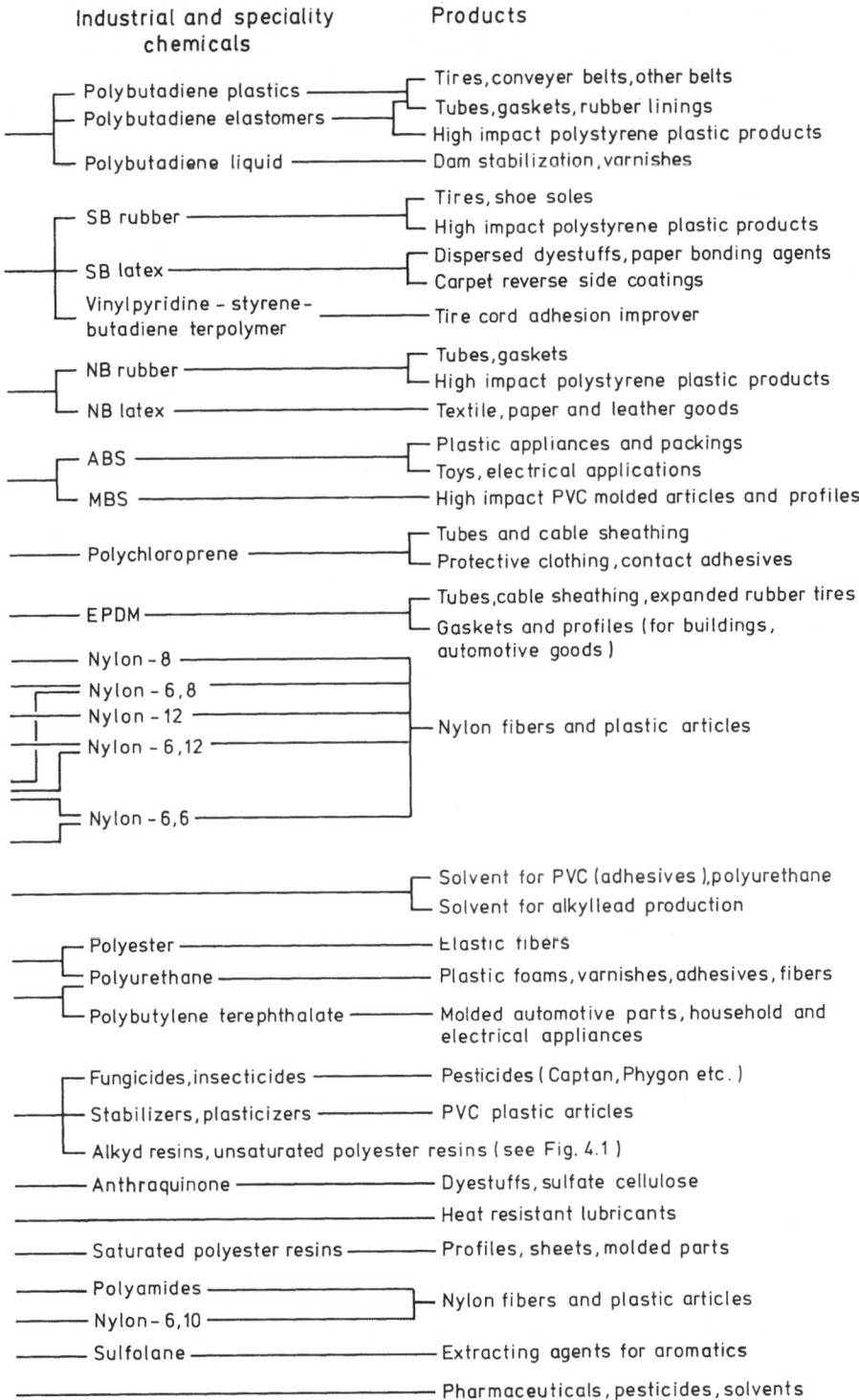

- Polybutadiene plastics —— Tires, conveyer belts, other belts
- Polybutadiene elastomers —— Tubes, gaskets, rubber linings
  - High impact polystyrene plastic products
- Polybutadiene liquid —— Dam stabilization, varnishes

- SB rubber —— Tires, shoe soles
  - High impact polystyrene plastic products
- SB latex —— Dispersed dyestuffs, paper bonding agents
  - Carpet reverse side coatings
- Vinylpyridine – styrene – butadiene terpolymer —— Tire cord adhesion improver

- NB rubber —— Tubes, gaskets
  - High impact polystyrene plastic products
- NB latex —— Textile, paper and leather goods

- ABS —— Plastic appliances and packings
  - Toys, electrical applications
- MBS —— High impact PVC molded articles and profiles

- Polychloroprene —— Tubes and cable sheathing
  - Protective clothing, contact adhesives

- EPDM —— Tubes, cable sheathing, expanded rubber tires
  - Gaskets and profiles (for buildings, automotive goods)

- Nylon – 8
- Nylon – 6,8
- Nylon – 12
- Nylon – 6,12 —— Nylon fibers and plastic articles

- Nylon – 6,6

- Solvent for PVC (adhesives), polyurethane
- Solvent for alkyllead production

- Polyester —— Elastic fibers
- Polyurethane —— Plastic foams, varnishes, adhesives, fibers
- Polybutylene terephthalate —— Molded automotive parts, household and electrical appliances

- Fungicides, insecticides —— Pesticides (Captan, Phygon etc.)
- Stabilizers, plasticizers —— PVC plastic articles
- Alkyd resins, unsaturated polyester resins (see Fig. 4.1)
- Anthraquinone —— Dyestuffs, sulfate cellulose
  - Heat resistant lubricants
- Saturated polyester resins —— Profiles, sheets, molded parts
- Polyamides —— Nylon fibers and plastic articles
- Nylon – 6,10
- Sulfolane —— Extracting agents for aromatics
  - Pharmaceuticals, pesticides, solvents

* other speciality secondary products, see Sect. 3.4.7
** secondary products not dealt with (see Sect. 3.4.2 for reason)

production statistics do not distinguish between EPM and EPDM and give no information about the diene components used. With a consumption coefficient of 0.03, the EPDM amount scarcely influences the requirement for butadiene in any case. The amounts quoted for the rubbers are for non-compounded types, i. e. they contain neither filler nor rubber chemicals and are not oil extended. The amounts of· latices are based on their dry weight. The high proportion of SB-latex is conspicuous and Western Europe production is well above the average.

Table 4.10. Amounts of butadiene derivatives produced in the world during 1984 (in 1000 t/a) [4.4]

Derivative	North America	South America	Western Europe	Eastern Europe	Rest of world	Total
Polybutadiene (BR)	405	45	230	215	230	1125
SB-rubber	1170	190	655	1225	620	3860
SB-latex	230	50	345	155	85	865
NB-rubber } NB-latex	70	5	90	75	55	295
ABS/MBS	360	10	400	10	220	1000
Polychloroprene (CR)	115	–	90	120	35	355
HMDA	215	–	95	–	–	310

4.4.3 Specific Butadiene Consumption Coefficients for the Secondary Products

Butadiene consumption coefficients for the most important sedondary products are listed in Table 4.11.

Table 4.11. Butadiene consumption coefficients t raw material/t product
The raw material is butadiene in each case

Product	Process	Coefficient
Polybutadiene (BR)	Bayer/Hüls	0.95
SB-rubber and -latex	Bayer/Hüls	0.58
HSBR-/HSBL-resins	Hüls	0.40
Carboxylated SB-latex	Hüls	0.45
SBS-/SB-vinylpyridine terpolymer	Shell	0.38
NB-rubber and -latex	Bayer	0.66
ABS/MBS	Bayer, BASF/Röhm	0.22
Polychloroprene (CR)	BP-Distillers	0.80
EPDM	Hüls	0.03
1,5-Cyclooctadiene	Hüls-, Toyo Soda-COD	1.20
1,5,9-Cyclododecatriene	Hüls-, Toyo Soda-CDT	1.20
HMDA	DuPont	0.63
Adipic acid	BASF	0.47
THF	Mitsubishi, BASF	0.83
1,4-Butanediol	Mitsubishi, BASF	0.67
Tetrahydrophthalic anhydride	Hüls	0.35
Sulfolane	Shell	0.45

4.4.4 Processing Scheme for Butadiene

Table 4.12 shows world uses of butadiene, classified according to secondary products and regions. About 56% of world butadiene production is employed for making SB-rubber and -latex. The columns "miscellaneous specialities" contain mainly the secondary products 1,4-butanediol/THF; COD/CDT; THPA; and sulfolane.

Table 4.12. World processing scheme or end use pattern for butadiene, 1984 (in 1000 t/a)

Derivate	North America	South America	Western Europe	Eastern Europe	Rest of world	Total
Polybutadiene (BR)	385	45	220	205	220	1075
SB-rubber	680	110	380	710	360	2240
SB-latex	135	30	200	90	50	505
NB-rubber } NB-latex }	45	5	60	50	35	195
ABS/MBS	80	5	90	5	50	230
Polychloroprene (CR)	95	–	70	95	30	290
HMDA	135	–	60	–	–	195
Miscellaneous specialities	40	–	60	–	35	135
Total	1595	195	1140	1155	780	4865

5 Competition Factors of the C$_4$-Hydrocarbons

5.1 Macroeconomic Competition Factors

5.1.1 Competition Criteria of Products in the Application Hierarchy

The politico-economic process of operational productive performance is initiated by the demand of the ultimate buyers of the products. The consumers determine the type and amount of the products to be supplied through the purchasing power available to them in combination with the order or priority of their needs.

The requirements of raw materials, auxiliaries, fuel and operating supplies [5.1] and of production equipment and services rendered in the production hierarchy are set in motion retrospectively by the needs of the ultimate buyer, via the endeavour of the producers to obtain a reasonable profit by participation in fulfilling these needs. Figure 5.1 shows the production and utilization hierarchy of chemical products from the primary materials to the end use pattern corresponding to the ultimate applications.

The production of chemicals follows on from the winning and processing of raw materials as primary products, outside the chemical industry. It begins with the production of base chemicals and leads via intermediates to the industrial and speciality chemicals. From these many different articles are produced by the non-chemical processing industries (e. g. plastic products, textiles, rubber products etc.).

Crude oil and natural gas are the raw materials of petrochemistry. They are processed by being split into fractions which, as feedstocks (e. g. naphtha, ethane) are mostly pyrolysed (cracked) into the base chemical, ethylene. The most important co-products of this first production stage are propene and the C$_4$-cut.

The quantitatively most important end products (plastics, fibers, elastomers, basic materials for lacquers and detergents) are yielded via several production steps of aliphatic organic chemistry.

Most chemical production processes can be divided into multistep sequences of intermediates production. Similarly, processing of the chemical end products outside the realm of the chemical industry can take place in several steps (giving semi-finished, semi-manufactured goods). More than one company can participate in these chemical production and mechanical processing stages. Each step of this production hierarchy is divided further into the individual conditions of procurement, production and marketing of the rival producers. In each case the particular criteria governing supply mirrors the criteria of demand of the following production stage and finally those of the ultimate buyer. A product can be successfully marketed only when the criteria of supply and demand are in harmony at every stage.

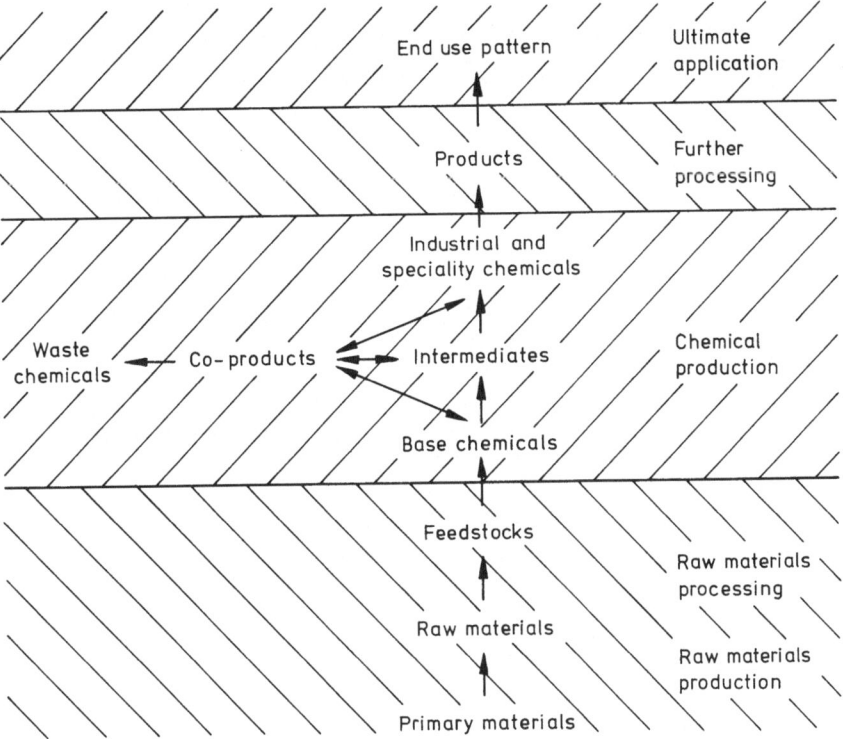

Fig. 5.1. Processing steps of chemical products

The chances of selling a product consequently depend in the end on whether the secondary products which can be made from it fulfil the technical and economic requirements of the end usage. Of interest here are first the technical-economic advantages at each production stage and then the competitiveness of the products in their individual end use fields.

Whether a product succeeds or not on the market is always decided in the final analysis at the level of the end use of the products made from it. Depending on how successfully these goods compete with others for similar purposes but made from different raw materials and/or using other processes, demand develops in accordance with the preceding operational performances in the production hierarchy. The competitive advantages which are gained in the preceding production stages are hence, in principle, only a means to the end of being able to prepare competitive end products.

Market studies of the prospects of success of chemical products must therefore anticipate as far as possible their conditions of supply and demand in the secondary markets. Table 5.1 shows an example, of a relatively rough classification into fields of application of ABS plastic products in W. Europe. This is decisive for the further sales chances for butadiene in the production of ABS.

Table 5.1. Consumption pattern of ABS in W. Europe, 1985

Processing/use	% of total consumption
Auto industry	25
Domestic electric appliances	20
Radio, TV, electronics, telephone	15
Refrigerators, packaging, etc.	13
Pipes	2
Miscellaneous (toys, furniture, etc.)	25
Total	100

5.1.2 Limiting Factors of Trading

The possibilities of supply and procurement of products between the individual production stages do not depend only on personal decisions of supplier and consumer, i.e. on the economic factors which can be influenced by these trading partners alone. On the contrary, the limiting factors in the markets are those which decide the intermediate and long term prospects of a product; these govern the superimposed supply and demand criteria and can be influenced little or not at all by the trading partners. These limiting factors are made up of a large number of single factors which can be classified into six main groups:

> macroeconomic
> ecological
> technological
> legal
> social and ethical
> political

Here and there in the literature, divergent opinions are to be found, e. g. "scarcely controllable influencing factors" to which are counted "economic system, political system, legal framework, sociological structure and cultural environment" [5.3]; or "environmental systems" with the subsystems "economic and technical systems, political, legal and regulatory systems, cultural and life-style systems, other social systems" [5.4].

Many extraneous factors of influence may be assigned to more than one of these main groups. Thus raw material potentials are a macroeconomic (e.g. state of resources) and also an ecological one (e.g. irreversible exploitation of resources). Ecological factors are mostly effective only in combination with political decisions and legal (legislative) limiting conditions. Adequately comprehending and classifying these extraneous factors are still less difficult than analysing and forecasting their mutual interactions, which is only possible through system-analytical considerations.

Collective action of individual trading partners can influence to some extent the limiting factors (e.g. pushing through requirements for environment protection or subsidies, establishing legally based quality standards). Some powerful suppliers or consumers have the capacity to exert influence, especially in the technological field (e. g. fixing technological standards or directives in the so-called state of technological progress or state of the art).

In the sphere of manufactured goods the influence of the limiting factors can be directed towards the conditions of procurement, production and marketing of products (product related) or towards the suppliers and consumers (company related). The product related (referring to market objects) action influences either the marketing prospects of all products competing with each other and of all their suppliers or the procurement possibilities of their consumers (e. g. rise or fall of price of petrochemical products depending on the dollar exchange rate); or it is directed towards a particular product or a special group of products (e. g. the ban on use of leaded gasoline). The company related (referring to market subjects) activity can influence the marketing prospects of all suppliers (e. g. effects on commerce from the state of the economy) or only certain suppliers (e. g. subsidies or tax relief for small businesses) or consumers (Fig. 5.3).

The production and sale of unleaded as against leaded gasoline in the FRG have been encouraged during recent times by tax concessions. The tax on unleaded gasoline has been reduced to an amount of several pfennig less per litre than for leaded gasoline. Part of this benefit is transmitted to the consumer via lower prices at the filling stations. The remainder can be used to subsidise the higher processing costs of the unleaded material in the refineries. In this way it was possible to increase the market proportion of unleaded gasoline in the total amount sold in the FRG up to the end of 1987 to approx. 36% (Fig. 5.2). This proportion is far lower in other W. European countries, e. g., less than 1% in Great Britain. For this reason the consumption of several secondary products of C_4-hydrocarbons (TBA, MTBE, etc.) has developed differently in the different countries.

The market-orientated managements of enterprises wish to avail themselves of the chances which arise from the limiting conditions, notably from alterations in them and, through this, to avoid the risks implied by them. There have been few references to the increasing importance of taking all limiting conditions into account in future

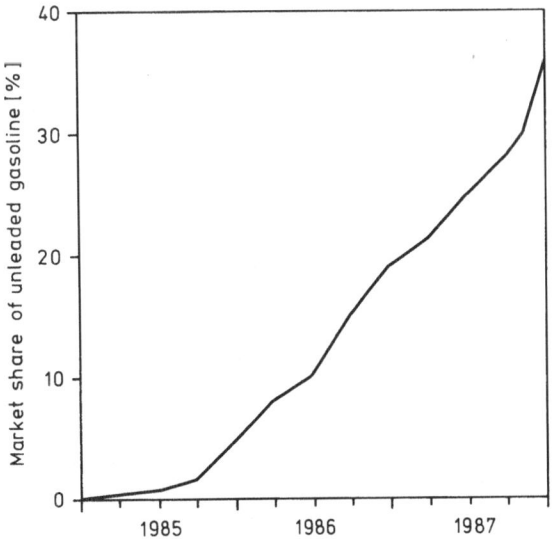

Fig. 5.2. Consumption of unleaded gasoline in the FRG [5.5]

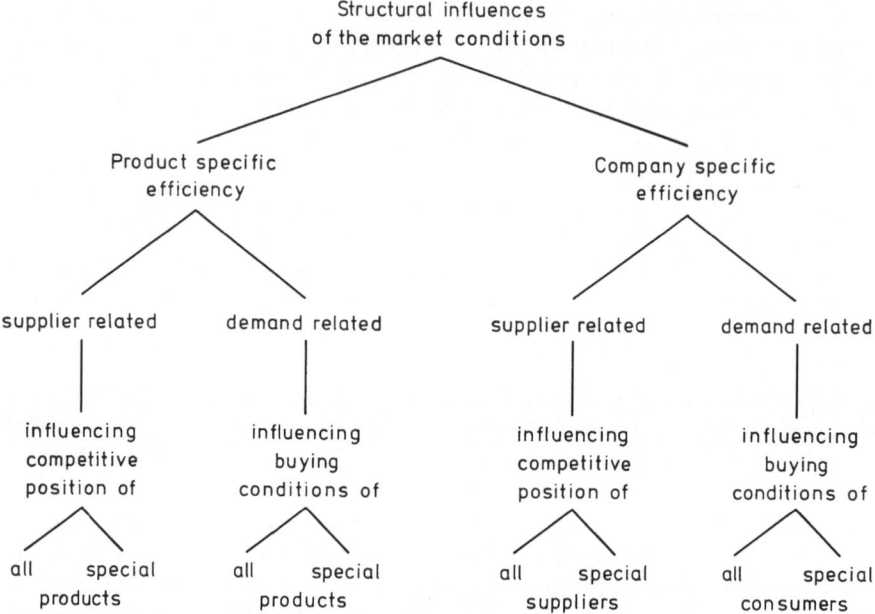

Fig. 5.3. Effectiveness of structural influences of market conditions on product and company

decisions about investment in the chemical industry [5.6; 5.7]. The central feature of the present book is the product related effectiveness in competition of the technological limiting factors.

5.1.3 Threshold Effects of Market Changes

The group of limiting factors in marketing affairs is continuously subject to alterations because new individual factors appear and others drop out and also because there are changes all the time in the effects of these single factors and in their collective influence.

  These changes may develop gradually (e. g. modifications of supply and demand due to changes in the business situation); or abruptly (e. g. political decisions or legal settlements). Many gradual changes of the limiting factors effect sudden changes in the competitive chances of a product when a particular threshold is reached which bestows or takes away this competitiveness. This is typical for competing products with, for example, differing labour costs. As these increase, the product with higher labour costs loses its competitive power beyond a certain point. The same applies to material and energy costs, also for successive improvements in quality in order to reach the level necessary for a product to compete in a new field of application.

  As an example, the competition between synthetic and natural rubber has been influenced by numerous factors. The falling petroleum prices, from the 1950s on, led to lower costs for the raw material monomers (butadiene, chloroprene, *i*-butene, etc.) and helped synthetic rubber with its various types of product to assume a commanding

position in the competition against natural rubber. Thus synthetic rubber has substi-
tuted natural rubber more and more. Figure 5.4 shows the developments in world
rubber consumption. The percentage participation of natural rubber was 60 in 1958
but had fallen to approx. 30% by 1980. The oil crises of the 1970s brought about
marked increases in price for synthetic rubber which gnawed at the price advantages
in the competition. The present market share of about 33% for natural rubber seems
to indicate that the substitution process has temporarily come to a standstill.

Technological innovations usually become recognizable only in stages, via scientific or technical progress, and then make their economic break-through suddenly as a
result of procedural or product perfection.

In the same way, legislation for protecting the environment or regulations for
admissibility of new products are rarely imposed unexpectedly. The decision-making
on such requirements is generally preceded by periods of discussion. When they
become law, they then have a sudden influence on the market prospects of the
particular products.

The intermediate- and long-term competitive abilities of the producers, and hence
also the total sales volume of their products, depend decisively on two superiorities
over their rivals — firstly their ability to recognise, more speedily and more far-
sightedly, possible changes in the limiting factors of the market and secondly their
ability to make a more reliable forecast of the effect of these changes on their
company- and product- related competitiveness. This enables them to adapt their
research and development activities better, and to increase the efficiency of their
procurement, production and marketing activities to the changes.

The marketing prospects of C_4-hydrocarbons depend, as far as these limiting
conditions are concerned, on several factors — utilization of technological innovations (production and processing of the C_4-hydrocarbons and their secondary products), adaption measures to ecological requirements (e. g. environment protection

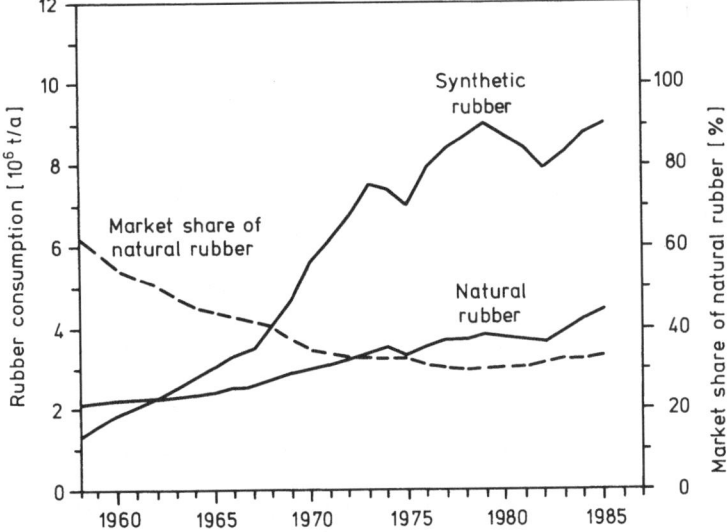

Fig. 5.4. Growth in world consumption of natural and synthetic rubbers

directives against lead alkyls in gasoline), macroeconomic factors (e. g. accessibility of raw materials through exploration of new petroleum or natural gas sources), and political decisions (e. g. OPEC politics). In the following pages dependence of former and future marketing chances of the C$_4$-hydrocarbons is treated, with particular attention devoted to the threshold effects arising from the limiting factors of the market.

5.2 Microeconomic Factors of Competition

5.2.1 Competition Factors Related to Products and Suppliers

The successful marketing of a product is the sum of the individual marketing efficiencies of its producers. The company-related competitive power is derived from its product-related and other marketing-related efficiencies. The product-related efficiency of a producer is an expression of how markedly he is superior or inferior to his competitors in his individual supply characteristics — quality, price and quantity of each product. Marketing-related efficiency is taken below as meaning how success- fully a supplier converts his product-related performance criteria together with his other sales-related performance criteria into individual marketing success.

Marketing-related efficiency is especially important for those suppliers who are obliged to have sales successes despite having the same product-related efficiency. Producers and also traders are in this position when they compete with each other with the same products obtained under approximately the same conditions of procurement and production. Their chances as suppliers of gaining supremacy in competition lie in excelling in other performance criteria of sales proficiency, such as superior sales and service efficiency, opening up new markets of application, better publicity (e. g. through advertisement), reliability of delivery, giving easier terms of payment, giving guarantees, reserve capital in the event of tortiabilities, image of the firm, its co- workers and products, etc.

In contrast to these more marketing-related competition factors, the differences in sales success of producing enterprises are usually due primarily to their product- related performance criteria. This applies particularly to the chemical industry. In exceptional cases the additional marketing-related know-how influences the general sales prospects of a product when all the suppliers of the product are superior in their collective marketing-related know-how to the suppliers of competing products of a different kind. Thus the producers of synthetic products owe their success in competi- tion with suppliers of natural products (e. g. synthetic versus natural rubber) clearly to the powerful marketing activity of the chemical industry. This is expressed in the special adaptation of the individual types of product (e. g. numerous types of synthetic rubber) to the particular needs of the buyers, the comprehensive customer service, relatively high consistency of quality of chemical products, reliable supply, and the compensating price policy which arises only feebly in the marked fluctuations of the base chemicals from natural materials.

Increase in the product-related competitiveness augments the general sales chan- ces of a product at the expense of competing products. The residual marketing-related ability of the suppliers, i. e. without the product-related competitive features, usually influences only the individual marketing chances of the single suppliers in that their sales successes increase their share of the market without altering the total sales opportunities of the product.

The potential for product-related competitiveness resulting from its conditions of production and resources, forms the basis for the general marketing prospects of a product — the technical properties, production costs, and the quantity produced (Fig. 5.5). The scope of the present book is limited to interest in the general marketing chances of the C_4-hydrocarbons. These product-related potentialities for competition are thus the centerpiece of the treatment.

5.2.2 Criteria of Quality as Factors in Competition

In the narrower sense of the expression the quality of a product means the criteria of its technical serviceability during use or consumption. In the wider interpretation it can comprise all the qualitative features of an offer — i. e. all features except the criteria of price and quantity. The interest of this book is devoted predominantly to product-related criteria of competition. Consequently, the term "quality" will generally refer only to the technical properties of the products.

The price or the amount of a product is a criterion of competition which is fixed in one dimension by a single information value. Its technical properties, however, are made up of a large number of individual criteria. In order to evaluate their competitive effectiveness they must be compared with the corresponding properties of the competing products. There is a vast number of property characteristics so that these must be subdivided. This can be done in three dimensions according to:

— effectiveness of the properties in the utilization hierarchy
— economic and metaeconomic effectiveness of the product properties
— the minimum properties necessary for a product to be able to attain success on the market

5.2.2.1 Classification of Product Properties According to the Application Hierarchy

Starting from the production of the crude material the properties of a product can be subdivided into:

> basic properties
> processing properties
> application properties

The basic properties comprise the data about its material composition (chemical, physical and technical properties, such as molecular structure and weight, particle size and structure, bulk density of solid materials, weight per litre; and viscosity features for liquid products, etc.)

The processing properties concern the workability and especially its advantages for processing through which a processor prefers it to other products. The advantages can involve savings of processing costs or manufacturing of products with superior properties.

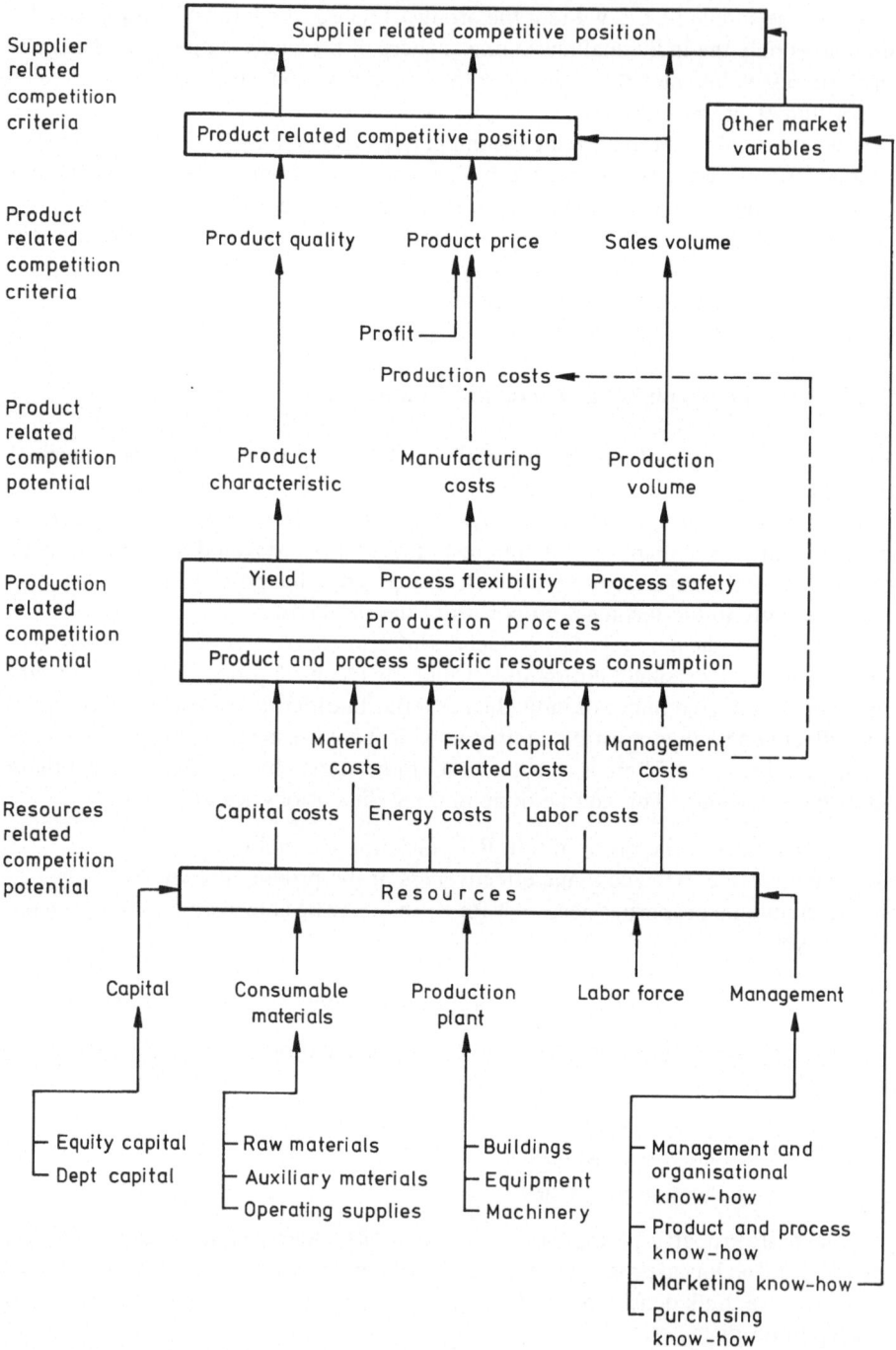

Fig. 5.5. Structure of competition criteria and potential

The utilization or application properties comprise the application properties of the end products which can be made from the particular product. There is a large number of end products (durable and consumable goods) and hence the end utilization properties of a product can be named only according to the special type of end product. The utilization properties of interest for end products from chemicals, i.e. plastics, synthetic elastomers, synthetic fibers, special materials of silicon chemistry are:

mechanical properties	(e.g. tensile strength, breaking strength, rigidity, resistance to abrasion)
electrical properties	(e.g. electrical conductivity, dipole properties)
chemical stability	(e.g. stability towards solvents)
safety and environmental technical data	(e.g. low inflammability of building elements)

The basic properties of a product are merely indirect criteria for judging its qualitative competitiveness. They are of interest because many of its processing and utilization properties can be derived from them. They are particularly informative when small improvements of a definite basic property can achieve a remarkable amplification of the processing and/or utilization properties of a product.

The processing and application properties of a raw material are due not only to the basic properties; a part is also played by auxiliary materials which are needed to prepare end products with definite properties. Figure 5.6 gives an example in which the toughness of plastics, ABS in this case, can be increased by reinforcement with glass fibers.

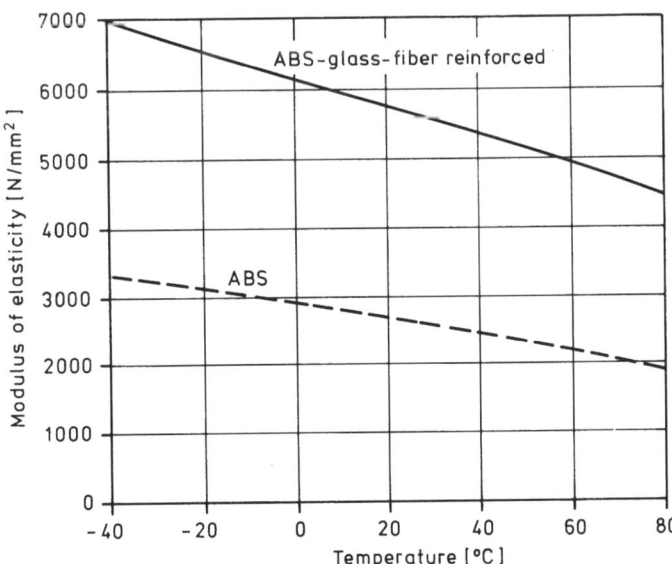

Fig. 5.6. Modulus of elasticity of glass-fiber reinforced ABS compared to normal ABS at different temperatures [5.6]

5.2.2.2 Economic and Metaeconomic Criteria for Quality

Furnishing a product with desirable quality criteria costs money. In particular, improvement of product properties usually leads to an increase in its production costs. Spending this money or extra money is worth while only when the quality benefits can be transmuted into profits via success in competition.

The increase in the competitive ability of a product is expressed through its advantageous qualities raising the economic benefits of its use because, at the processing stage, improvement of raw material quality leads to lowering of processing costs, and/or because performance chemicals of higher quality can be made from the better raw material.

The producers at the forefront of the production chain endeavour to participate in this increased profit of the processors and/or the ultimate buyers by expecting that their gross profit on sales will rise out of proportion to the increase in their costs. This can be done either directly by increasing their prices, or indirectly by maintaining prices unchanged and increasing sales.

However, the economic profit from utilization expressed in the cost/benefit ratio, is not the only factor determining the competitive power of a product. Acceptance or rejection of a product can, in addition, be decided by properties which cannot be evaluated economically. These are henceforth termed "metaeconomic" properties.

Properties of products and features of processes which can be evaluated economically may be compared with each other objectively by relating prices or costs (see Chap. 6). In contrast, the influence of metaeconomic properties on the competitive position of a product can be estimated only subjectively. It is thus difficult to do more than speculate over the market chances of products where supplier and buyer have a free choice between products which are ecologically dubious but more economically usable and those ecologically less questionable but more expensive (e. g. the market chances for unleaded gasoline if it were more expensive than leaded).

5.2.2.3 Property Standards of Products

The basic, processing and/or application properties of many products are fixed through standards. Minimum property requirements are prescribed for the raw materials or the semi-finished or semi-manufactured goods prepared from them. These requirements should be displayed or there is a legal obligation to do so.

The simplest standards are terms of acceptance from buyers. Thus car factories quote minimum technical properties which their suppliers, of tires for example, for the initial standard equipment have to fulfil. They are faced with the technical supply conditions with which, for example, tire manufacturers can make their products comparable for the trade in tires and the end consumer (replacement requirement for tires).

Legal demands of minimum properties for products are appreciably stricter. They exist in all industrial countries for product applications and production processes in which it is imperative to adhere to definite criteria of quality in order to prevent personal injury and damage to property. This is the case in, for example, civil engineering (statical properties and degrees of combustibility of standard building elements used for the internal outfitting of rooms), the transport sector (methods of transport and their infrastructure), the foodstuff field (foods, packaging materials)

protective labour legislation and disposal of products (non-return packing, conservation of clean air, water and ground).

The regulations for production, supply and usage refer to particular products or
they are issued as global conditions which do not refer to the use of a definite product
or compound class.

The values of many properties can vary according to the method of their
determination. Standard requirements are therefore usually coupled with the relevant test conditions. The standard minimum properties which a product has to have to
be acceptable considerably reduces the versatility of the task of basing the competitiveness of products by mutual comparison of their mostly multiple features of quality.
This simplifies, above all, analyses and forecasts of sales price acceptability decisively
because many of the products included in the comparison of competitiveness fail to
surmount the obstacle of demanded minimum properties.

5.2.3 Prices as Factors of Competition

5.2.3.1 Prices After Fulfilment of Standards of Minimum Properties

In many sectors of the market, demands are merely made for fulfilment of minimum
technical requirements which are fixed by standardised conditions of acceptance. The
buyers are unwilling to pay the higher prices if fulfilment of properties is carried to a
higher level than this. The failure to respect a minimum requirement cannot be
compensated by overfulfilment of another. Further, such a shortcoming cannot be
compensated for by price reduction.

From a business-political point of view the big purchasers of products are rarely
content with a single supplier (risk of failure of a supplier through breakdown, strike,
bankruptcy etc.). Hence not just the supplier with the most favourable price is
accepted. The next best and perhaps further suppliers in the sequence of their prices
quoted, come into consideration, depending on the demand for the product and the
production capacities of the suppliers. Decisive for acceptance of the quoted products
is the magnitude of the net prices, after deduction of discounts, bonuses, etc. at the
site of processing by the consumers.

5.2.3.2 Economically Calculable Application Properties of a Product
as Price Parameter

If the product consumer is a processor wishing to make end products from it, he is
interested in two types of economic advantages of the product offered to him:

1) Savings in costs, which can be achieved compared to processing alternative
   products which are available (processing cost advantages).
2) Possibility of preparing secondary products with better application properties than
   those of secondary products which can be made from competing products (advantages of end use properties).

Concerning 1) Savings in processing costs yield direct increases in profit for the
producer by lowering the production cost. By producing a secondary product with
better application properties he can increase profits by price and/or sales increases.

Both types of profit improvement are effected through economic advantages of the better product — through more economic processing, or better end application properties of the products which can be made from it.

Lower processing costs of a product are derived via:
— saving in material costs, e. g. through
  — preparation of thinner-walled articles
  — reducing the amount of waste
— saving in energy costs, e. g. through
  — lower requirement of energy for machine operation in the plants for product processing
  — lower requirement of heat energy for product processing
— saving of labour costs, e. g. through
  — simpler processing conditions
  — higher throughput of product
  — lower susceptibility of the processing installation to repair
— saving of capital related costs, e. g. through
  — higher output of product
  — easier product processing so that there is less wear and tear of the processing equipment.

These cost savings must be made clear in order to show the product processor the price level up to which it is more economical for him to prefer an improved or new product.

Concerning 2) The efficiency of the end application properties of a product is most simply evaluated economically by deriving performance prices from the properties and prices. These show whether the application advantages of a better product outweigh its higher price, compensate for it or do not justify it. A price comparison of this sort includes physical properties, such as mechanical properties (e. g. price per 10 N tensile strength), electrical properties (e. g. price of a cable insulation based on the insulating power) or thermal insulating power (e. g. price of the insulating material per kJ/h and m^2 heat insulation).

Instead of physical properties, the production costs or prices can be based on the performance value of the finished products, e. g. prices per metre pipe of plastic, iron or copper referred to a standard cross-section and pressure stage, the requirement of bottles or tins of tin-plated iron, alumina or plastic per litre charge converted into price/kg or price/dm^3 material.

It is often the case with auxiliary chemicals that a new product may have higher production costs but, compared to a product hitherto in use, has an increased effectiveness out of proportion to this higher expense. Here, too, evaluation of its economic value demands a weight-price comparison. In this, it must first be ascertained how much of the competing materials is needed to attain the same efficacy; these then have to be converted into convenient price units.

Economic comparisons of this sort are not complicated provided only pairs of data are taken which are comparable in their dimensions. If several criteria of efficiency have to be related to each other, multiple dependences can arise. For example, the competitiveness of anti-knock gasoline additives does not depend alone on their varying suitability, weight for weight for increasing octane number, but their differing energy contents must also be considered. The wider the property spectrum of a product, the more extensively must the comparisons of efficiency be performed.

5.2.3.3 Price Evaluation of the Metaeconomic Benefits of Use

Metaeconomic properties are regarded differently by industrial consumers who process the product further than by those who are private end users. Private users judge the good value of most products primarily from the point of view of the metaeconomic utilization. Only after this do they choose the cheaper or more economical among equally suitable alternative products.

This does not mean that producers judge the value of a product purely on economic criteria while ignoring metaeconomic features. They, too, are ready to acknowledge by paying for the metaeconomic advantages of a product (prestige value, safety properties, ecological advantages) and to do this sometimes even with disregard of economic criteria.

The difference between these two types of consumer is that the private user views product offers with appreciably more emotion than the producers; their scale of value criteria includes, in addition, considerably more exclusively metaeconomically assessable properties.

Most of the metaeconomic properties of a product can indeed be measured objectively. The profit from this for the product user is, however, assessable only emotionally. The profit criteria which are derived from this can be apportioned only within a classification grid from which limited gradations of value but no price forecasts can be made.

Most product properties which can be utilized metaeconomically may be subdivided into:

1) Properties which promote the well-being of the product user
   a. bodily satisfaction, e.g. comfort (from floor heating — where poly-1-butene tubing may be used); ease, health (use of TBA or MTBE instead of lead alkyls for increasing octane number); ergonomic arrangement of equipment at place of work
   b. mental satisfaction, e.g. appealing shape and colour of products (such as use of galvanised ABS-plastics), gratification of desire for recognition (employment of transparent PMMA panels in imposing buildings), appraisal of feelings of speed and acceleration at the car wheel (use of fuels of high octane number), safety, mobility
2) Properties which endanger health and well-being
   a. directly dangerous, e.g. from poisonous, inflammable, or explosive materials; tendency to cause accidents, causing organic damage
   b. ecologically hazardous, e.g. dissipation of resources, significant impairment of viability of plants and animals through production, use or disposal of a product
3) Properties influencing well-being and perceived via the senses
   a. optical properties (transparency to light, colour, shape, design)
   b. acoustic properties (resonance, sound attenuation)
   c. properties of odour and taste, allergy provokers
   d. properties perceived by feeling and the sense of touch (cold, warmth, hardness, softness)

The danger and ecological jeopardy associated with products can, in contrast, be perceived only by instinct and experience, and their action judged only by standards of reason.

Above all, when taking into consideration the social components of the readiness to pay the price, a distinction must be made between two benefits:

1) The metaeconomic benefit which comes alone and directly to the product user, e. g. a prestige gain, or a benefit which they enjoy together with others, e. g. using unleaded gasoline.
2) The profit which they themselves do not have but which is enjoyed by others, e. g. investment of an employer for improving working conditions for his employees.

The metaeconomic properties of products become effective typically in two spheres of application, and often even displace wholly economic criteria:

— The danger of a product manifests itself as doubt about the human and ecological effects of its preparation, processing, use and/or disposal.
— End users, and often even producers, permit themselves to be influenced in their purchase considerations by metaeconomic product properties which are suitable for satisfying their ethical claims to prestige and recognition.

It is important for decisions on investment in favour of such products to know how highly the metaeconomic properties are acknowledged financially by the consumers and/or through increase of the market volume.

5.2.4 Quantity as a Factor in Competition

The competitiveness, especially of those companies producing industrial goods, depends firmly, for various reasons, on the size of the amounts offered:
— The companies are accepted as suppliers only when they are able to supply the particular product continuously and in amounts which do not fall below a certain minimum. In a market where large amounts of a product are being processed, a supplier has no chance when he can only deliver insufficient quantities of a product, even if it is of superior quality and/or extremely low-priced. In the meantime numerous chemical processes have been developed so that alternative possibilities prevent bottlenecks in the supply of individual base products; holdups in production can then be avoided.
— New chemical products, developed to replace others, have first to be produced on the laboratory, and then at the most, on the pilot plant scale. The producers then rely on their customers' rendering a service by processing experimentally small amounts of a new product-which will be provided regularly in adequate amounts only some years later. This helps the supplier estimate the usefulness which offers the chance, gradually over months or years, of becoming established as an additional supplier.
— The larger the amounts which a supplier can produce in relation to his competitors, the greater are the cost advantages from the economies of scale, and the more likely he is to become the market leader in price policy.
— The larger the amount produced, the greater is generally the constancy of product

quality. Inversely, smaller suppliers, who want to become established in smaller plants as competitors, often suffer from the marked quality fluctuations of their products.
— As production capacity rises, the production costs of the suppliers fall and hence the chances of a price reduction of the product are better. The marketing of many products fails because the demand is insufficient to permit them to produce in quantities which keep production costs below the break-even point.

The potential buyer is also interested in another aspect of quantity as a factor in competition – whether increased demand will lead to shortages, regarded as an omen of price increases. It is further of interest to see with what price elasticity shortages of base chemicals react to increases in demand.

5.3 Costs as a Basis of Price Competitiveness of a Product

In order to ascertain whether a product is sufficiently competitive in price over a long period, or if it is better to improve production or even risk new development, the future costs of this product are compared with those of competing products, bearing in mind their differences in quality. Only the product with the more favourable costs can survive competition successfully in the long run. This cost-relevant condition for the price competitiveness of a product is henceforth termed cost potential for price competitiveness, abbreviated as "cost-relevant competitiveness".

The differences in cost of competing products are above all due to differences in production costs. It is therefore usually enough to reduce the cost comparison to that of production costs. Only when marketing factors specific to the business or product have a special significance for competitive ability (e. g. customer service, special sales strategies) does the cost-relevant competitiveness have to be evaluated on the basis of the full costs, including marketing, administration and R and D costs.

5.3.1 Cost Reduction via Increased Sales

5.3.1.1 Lowering the Procurement Costs

A producer is able to achieve price or cost reductions in the procurement of base chemicals in relation to competitors in three ways, namely by:

1) cheaper purchase of the base chemicals
2) own production of the base chemicals
3) purchase or own use of co-products at the same time.

Concerning 1) Considerable discounts for quantity are obtained in commercial dealings with base and intermediate chemicals. A producer making a product in amounts appreciably larger than a rival, hence having a much greater requirement for the base chemical, can benefit from large price reductions. Since the base chemical costs in the chemical industry amount to 50 to 80% of the production costs, the production costs are decisively influenced by such discounts.

The prices of base and intermediate chemicals fluctuate in accordance with changes in the supply and demand situation much more than those of secondary products. Temporary price advantages can be attained by optimum procurement logistics. Price peaks in boom periods can be countered somewhat by long-term supply agreements — but this often prevents advantage being taken of lower prices in the spot market during slump phases. Prices in the chemical industry are fixed for periods of only one to three months. The prices develop differently in various countries so that large firms which purchase internationally and especially if they have their own acquired knowledge of world markets and technologies, are able to buffer these price fluctuations and use them to their own advantage better than can the smaller consumers.

Alterations in the procurement market are due to three superimposed causes:
— Changes in the supply and demand situation are brought about particularly by macro-economic factors of alteration in the state or structure of the economy.
— Technological innovations lead to an increased demand for the base chemicals necessary for new processes and cause the demand for the base chemicals for older processes to decrease.
— Reasons specific to the firms (e. g. breakdowns, revamps or maintenance of a production plant) bring about occasional disruptions in production. This leads to a decrease in demand for the base chemicals and to a shortage or price increase of the materials prepared from them.

Concerning 2) If there is an intermediate and long-term shortage of base chemicals and especially of raw materials it helps to evade imminent price increases by participating in a raw material business. This is, however, a disadvantage if the spot price falls below the production costs of the product as a result of the supply becoming normal again. This has occasionally been the case in the mineral oil sector.

Concerning 3) The price advantages are especially good when the particular base chemical is produced along with a co-product and this co-product complex can be exploited more profitably than by a supplier who has had to search elsewhere for sales possibilities for this co-product. Thus the entire C$_4$-cuts from steam cracking plants are often purchased and so all the possibly useful products are obtained. This permits the use of larger and cheaper processing capacities for the C$_4$-cuts. Such a powerful situation of product utilization suggests therefore taking over the production of base chemicals in the preceding stage.

A backward integration is not worth while, however, when the base chemical is yielded as a co-product which may possibly be difficult to market and is therefore valued at only a nominal sum and disposed of at any price.

The widespread exploitation of co-products has given the industrial nations a marked competitive advantage over the Arabian oil-producing countries, in whose petrochemical complexes the co-products have hitherto been processed to only a limited extent.

5.3.1.2 Lowering the Production Costs

Production costs diminish with increase in production capacity and this decrease is larger, the greater the degression of the investment costs as a function of the capacity. This can be expressed approximately for chemical plants as a power function with an average degression exponent of 0.65 to 0.70. In addition, the production costs of the products fall with increasing degree of utilization of a particular capacity, whereby the lowering of costs is correspondingly more marked as the proportion of fixed costs increases. The lowering of the various costs per production unit is:

Capital-dependent costs in accordance with the degression of the investment costs. The most important of these are the depreciation costs which depend on the invested capital, the calculable interest rate, the maintenance costs and the taxes and insurances on the capital.

The *labour costs,* because the number of personnel for running a plant increases less than proportionally to the capacity, i. e. large installations require little additional service staff than smaller ones. The labour costs are radically reduced by a high degree of automatization. The larger a plant, the more worthwhile is equipment for measuring and special automatic controls.

The *material costs* with increase in the quantities produced and in industrial know-how, because of the reduction in the proportion of substandard charges or in the amount of low quality fore-running and discharge as production increases.

The *energy costs,* because the requirement of heating, cooling and driving energy can be more exactly optimized in large plants and is less for larger machines and apparatus.

The dependence of the savings in cost on the capacity of a plant can be recognised from the cost curve of the particular process or product. The more pronounced the cost degression is, the greater is the interest in deriving cost advantages by expanding production. All the same, the process with the most marked cost degression is not necessarily the permanently most cost-favourable. On the contrary, the costs of processes which are especially economic at full utilization often increase progressively when the capacity is not fully utilized, e. g. owing to reduced sales as a result of the state of the economy. The capacity-dependent cost degression is thus optimally usable up to that capacity at which the marketing can take care of sufficiently continuous sales and hence production amounts to meet the nameplate capacity and high degree of utilization of the production plant.

In order to be able to profit from the advantages of larger capacity the usual marketing strategy, anticipating the lowering of costs, coming from a larger plant with sufficient utilization, is to lower prices and achieve rapid increases in sales which then leads to the growth in profit which was the object of the exercise (so-called penetration strategy).

On the other hand, smaller competitors can also avail themselves of these capacity-influenced cost reductions, even through their individual sales chances are lower, by uniting to form a consortium or investment companies (Table 5.2). They can then produce together in one large plant instead of in several small ones, but each sells the product in his own way. If, later, when sales opportunities have increased, it becomes worth while for the individual producers to build their own plants of optimum size, the union can be dissolved and each possesses the production know-how they acquired together.

Production costs are also favourable when newly developed products can be made in plants which had previously been used for preparing other products, especially when the plants are no longer or only intermittently used. Thus the introduction of

Table 5.2. Examples of joint ventures and investment companies of C$_4$-chemistry in Western Europe

Country	Company	Site	Partners
France	Butachimie	Chalampé	Du Pont 50%, Rhône Poulenc 50%
	Distugil	Champagnier	BP 50%, Plastogil 25%, Rhône-Alpes 15%, ERAP 10%
	SOCABU	Port Jerome	Esso Chimie 80%, ATO 20%
FRG	Erdölchemie	Cologne-Worringen	Bayer 50%, BP 50%
	ROW	Wesseling	Shell 50%, BASF 50%

polybutylene terephthalate gained from the fact that it could be produced in the plants for polyester fibres. Many polyester plants closed down at the end of the 1970s as a result of the ruinous competition in the fibre market.

5.3.1.3 Lowering Marketing Costs

Companies active in international markets can delegate the selling of their products abroad to external sales companies. These can be requested by several companies at the same time to market their – non competing – products. Occasionally the marketing bureau of a single firm abroad acts also for other firms (e. g. the Bayer agents in Mexico take over selling of non-competing products of Hüls). Marketing logistic costs can be lowered when competing suppliers in their decentralised plants furnish the customers of their competitors who are in their vicinities. A typical example is the so-called swap activity, when different gasoline producers supply from their refineries neighbouring service stations of their competitors – but at the cost of, and using the composition formulas of the competitors. Thus the Esso refinery in Ingolstadt, the BP refinery in Vohburg and the Marathon refinery in Burghausen simultaneously meet the requirements of the Shell and other service stations in Bavaria. The gasoline additives, the nature and quantities of which distinguish the various gasoline brands from each other, are also supplied in accordance with the "prescriptions".

5.3.1.4 Recovery of R and D Costs

Producers have to carry out expensive R and D in order to be able to derive economic use from new scientific and technical knowledge. The innovations are then translated into new types of production processes, yielding improved or new products with properties suitable for marketing. The costs when products require intensive R and D can exceed by far the construction costs for the production plant, leading to higher full costs impairing competition decisively. Preparatory marketing activities are also needed before launching the regular sale of a new product, e. g. marketing research work, programmes of market development, training of technical service and sales personnel.

R and D and marketing activities continue after the product has been brought onto the market. From this point the cost of this further upgrading of processes and products, and of improved sales performance belongs to the normal production and marketing costs of the product. On the other hand, the R and D and preliminary marketing performance costs, which have to be sustained at a high level for several years, especially for technical innovations, are added via the cost calculation to the cost of the products already in the sales programme. This is necessary because a large part of the R and D activities does not lead to market success and therefore their cost must be subtracted as a risk charge from the gross profits within the general firm venture. Apart from the calculable tax relief on these preliminary costs, they are an investment which when the risks of an enterprise are considered should be paid off as quickly as possible from profits. Two primary business activities can play an additional role here:

1) The engineering superiority can be utilized additionally in foreign markets by:
— exporting the product made by the particular process
— building abroad one's own production plants which utilize the process

2) Licensing the process yields fees which reduce preliminary expenses and give another opportunity of obtaining contributions to cover the costs.

5.3.2 Lowering Costs by Utilization of Technological Progress

5.3.2.1 Cost Reduction by Routine Work and Automatization

Every new production process, just as each essential technological change of a process already being carried out, creates starting or adaptation problems which are detrimental to the cost account as a result of slowed or disrupted production, unacceptably high amounts of sub-standard products, fluctuations in the quality of the products, and also increased consumption of raw materials, auxiliaries, fuel and active components. The additional costs incurred by these problems of running-in and adaptation become less as acquaintance with the special peculiarities of a process grows. Production experience with complicated processes continues to increase even after this running-in phase has been passed. This gain in experience, which can last for years, allows the risks of the procedure to be more precisely calculated (which was at first difficult or impossible) and thus eases the cost burden. It shows up notably also in an increase in productivity [5.7].

The increasing familiarity with new production processes usually ends in automatization steps, in which manual supervision and control are being replaced by fully automatic devices for measurement and control. This replaces labour costs by capital costs and enables production costs to be reduced over a period.

5.3.2.2 Cost Reduction by Process Changes

It is often possible, especially in the production of chemicals, to make a product, already marketable, in a more economic way using a new process or another, cheaper, starting material. In combination with the capacity-dependent decreasing trend in costs, explained in Sect 5.3.1.2, this gives three alternative ways of lowering the production costs of a particular material:

	Base chemical	Process	End product
1)	another	same	same
2)	same	another	same
3)	another	another	same

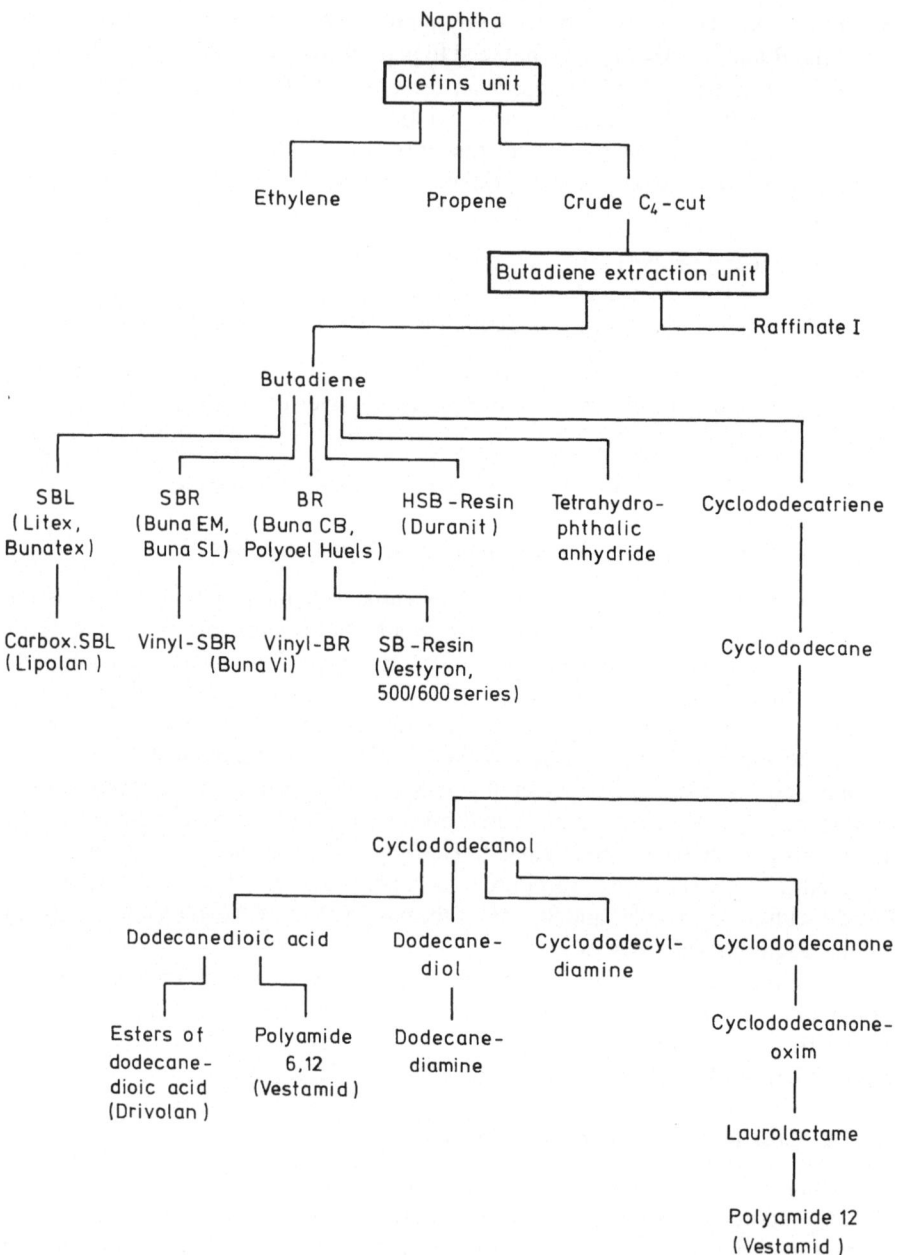

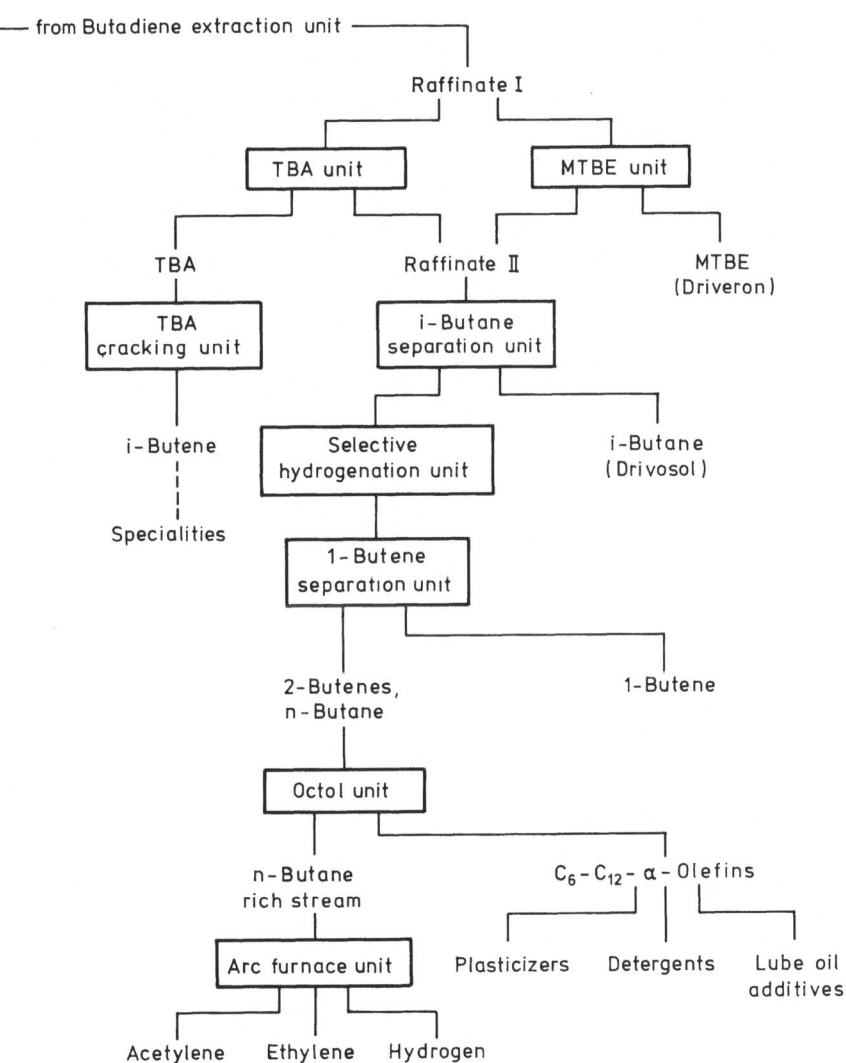

Fig. 5.7. Production scheme of C_4-chemistry at Hüls, Marl

If it is desired to produce a product more economically and yet retain the usual starting materials used and also the production process, this can be done only by utilizing the economics of scale, already described above. The production costs can also be lowered when it is possible to employ another, cheaper base chemical or another, cheaper process. The use of another base chemical usually requires changes in the production process and, vice versa, another process requires other base chemicals. The third variant (another base chemical + another process) is, therefore, the most usual way, together with utilizing economies of scale for lowering production costs.

5.3.2.3 Cost Advantages Through Production Integration

Non-chemical products are generally made in an individual straight sequence of production steps but chemical products, especially in the large chemical companies, are more usually made in historically developed vertical integrated production systems. These arose in a special way, specific to the firms and continue largely within a firm (internal combine) vertical combine economy, or with outside organisations (external combine) [5.8; 5.9]. This permits multiple use of basic products, and, above all, an opportunity of using co-products as base chemicals for cheaper manufacturing of products already known, or even of new products.

In this way, vertical integrated systems develop over many years and enable the unavoidable co-products to be utilized more effectively. These combined production systems contain historically optimised systems of both self-supply and disposal which develop horizontally and vertically in continuous adaptation to structural changes in the base chemicals supply and in the demand for end products, and to the technological progress of production procedures. These integrated systems have the economic and ecological goals of attaining the best possible overall material balance from raw materials to optimum economic usage of all side-products [5.9].

Figure 5.7 shows the production system of the vertical production hierarchy of Hüls, with the integrated C$_4$-chemical routes.

5.3.2.4 Cost Increases or Decreases Through Co-products

Changes to cheaper base chemicals and/or to a cheaper process technique yield cost advantages for processes which are accompanied by unavoidably large amounts of co-product, only when the co-products can be adequately exploited. A particular end product can often be produced more economically when an appreciably cheaper base chemical is used in the main line of a process. On further consideration of the co-product problem, the process nevertheless turns out to be uneconomic. This might be because the co-product is not sufficiently marketable and hence brings no or only a small cost credit contribution; or because the co-product is harmful to the environment and, through necessary disposal costs, increases production costs. Even a co-product for which there is a wide market can be uninteresting for its producer if it cannot be sold in sufficient quantities because of too high price concessions or on account of business-political sensitivities.

In such cases it must first be ascertained whether a change of catalyst or variation of the conditions can replace a rigid by a more elastic preparation of co-products, adaptable to the changing market conditions. If this does not suffice consideration must be given to technical possibilities of changing the co-products into marketable products or to basic alterations of the production.

5.3.3 Full Costs as Price Parameters

5.3.3.1 Problems of Evaluating Full Costs of Co-products

The C_4-hydrocarbons from both steam crackers and refineries are yielded as co-products with mainly ethylene and propene from steam crackers and with various liquid and gaseous refinery intermediates from refineries. The further processing of the C_4-hydrocarbons begins in both cases with the separation of the C_4-mixture into its economically exploitable components. The C_4-hydrocarbons produced using alternative routes via other production processes are similarly mostly yielded as co-products.

The producer of a material can take successful price policy decisions only when he knows the full costs of his products. This is the sole way in which he can recognise how far he will sell a product with profit or perhaps in contrast to his rivals – at a loss when he accepts a price.

Full costs of a product can be evaluated unambiguously only when the producer makes one single product. As soon as there are several products, even if they are produced in separate plants, the problem of the correct calculation and allocation of overhead and fixed costs (proportionalisation of the operating stand-by charges) begins. Calculation of the product-related allocation of costs of co-products is still less precise, because, in addition to the objective allocation of the sales and administrative overhead costs, the fixed and working capital costs and even the costs of material, energy and labour, can for the most part, not be broken down and accurately attributed. This is not even possible in retrospective calculations let alone in cost estimates made beforehand.

All the suggested methods for calculating full costs depend on the help of allocation schemes, which permit a more or less imprecise differentiation of costs according to the originating factors [5.10]. With these the costs are ascribed to the individual products partly using pseudo causal technical criteria and partly via chargeability according to the cost-bearing abilities of the co-products related to sales prices. The second dimension of these methods is that either the total costs of the co-product complex are directly partitioned between the cost-bearers, using lump sum clearing ratios (full-cost calculation) or by separating the direct and variable overhead costs from total costs which, according to causal criteria, is chargeable to individual cost bearers (direct costing). The remaining costs (fixed overhead costs) have to be carried by the main product only (residual value method) or they are distributed over all co-products, again with the help of an allocation scheme. As units for the cost distribution, the following suggestions have been made [5.11]:

— physical properties, e. g. mass, molecular weight, calorific value, density
— market prices

— utilization surpluses, i.e. sales prices of secondary products after deduction of further processing costs and marketing costs
— marginal costs
— costs of alternative production with the aid of competing one product processes

By combining the marginal and full cost calculations the fixed costs calculation is obtained [5.12]. It is the counterpart to the cost/credit-contribution analysis [5.13]. It starts with the assumption that all costs can be classified via definite references as special "cost packages" which are individually ascribed, like single costs, to the cost bearers.

In this way an attempt is made to obtain a target contribution margin for each co-product which gives the contribution that each product must make to cover the total costs caused by all products of the particular co-product complex together, plus the attained profit aimed at per accounting period.

The saleable amounts, and hence the total costs are dependent on capacity, and also the attainable prices, can be forecast with only limited reliability because they fluctuate appreciably over the long-, intermediate- and even short-term. Sometimes the prices and amounts sold of products related by their chemical production process develop markedly in opposite directions.

None of these costing methods, not even the break-even analysis, solves the problem of evaluating objective full costs for co-products. In contrast to its importance for producers with separate product lines, costing is therefore unsuitable as a device for planning and control for producers of co-products [5.14]. On the other hand, direct costing furnishes, in chemistry at least, a vague guidance, much better than having to take investment and marketing decisions completely without orientation using an unsuitable full cost calculation method [5.15].

5.3.3.2 Company Related Interpretations of Full Costs for Co-products

The problems and complexity and especially the low reliability of the methods for allocating all types of cost to the full costs of each product show that the lower limit of the lowest price of a product for breaking even does not depend alone on the different technological efficiencies of the production procedures; a decisive role is also played by the type of costing preferred and especially their way of cost allocation. It is thus quite common that two competing suppliers may base their price policies on different full costs or production costs even though they make their products from the same raw materials and use technically identical conditions for the processes in their plants.

The most advantageous conditions arise in this costing for the producer who is the first to succeed in utilizing economically a co-product, hitherto regarded as a waste product in the production of a single main product. He tends to burden the main product with the total costs of the co-product line, and to regard as profit the proceeds from the co-product or the secondary product made from it, minus the costs of the production-specific purification and marketing created by internal, further processing of the co-product. This profit is credited to the main product or, more likely, used to pay off as fast as possible the accrued costs needed for opening up the possibilities of utilizing the side-product.

As soon as this sort of exploitation of a side-product is also carried out by competitors, the prices of the main product will fall in accordance with the former profit contribution of the co-product, and these extra benefits can contribute increasingly only to covering the costs of the main product. This cost-regulating orientation of the profit is also the basis of the direct costing method.

5.3.3.3 Intercompany Transfer Prices

Products which are further processed only internally, within the company, have no market prices and gain no earnings, in contrast to those offered for sale externally. Instead the cost burdens are allocated to the following cost centres through transfer prices [5.16]. Intercompany transfer prices conceal many sorts of problems in the chemical industry [5.17]. If the transfer prices are fixed at production costs, losses in profit are generally suffered by those departments which produce relatively high proportions of base chemicals in the company; or, in exceptional cases, they may be protected from losses provided these base chemicals are available on the market at below production costs.

When transfer prices are derived from market prices there is scope for variation of the evaluation in the form of special conditions (rebates, bonuses, discounts), costs for packaging, transport, marketing expenses, commissions and suitable profit margins, which have to be negotiated between the departments or, in the event of a conflict, settled by the intervention of top management.

As a result of the internal obligation to accept intermediate products, there may be distortions of the competition which conceal the real contribution of the profit of individual departments to the overall success of the company.

In order to be able to make use of the direct cost method with the complex of co-products it is customary, as a result of the deficit in earnings, to utilize other substitute measures:

— If the self-produced product, further processed internally, is offered freely on the market, it has a market price. This can be used as a basis to establish the internal transfer price and for the fictitious proceeds of the direct costing analysis (net back minor value).
— If a co-product can be prepared by another process (if possible, in a single product line) the usually higher production costs for this can be calculated separately and used as a basis for the evaluation according to the direct costing method (production cost plus value).
— If it is an absolutely new product it has no market price, nor can it be made by any other process. It is then possible to obtain a fictitious market price from the prospective use which it offers to its consumers (production cost plus value).

5.3.4 Methods of Cost Estimation

5.3.4.1 Estimation of Production Costs

The production costs constitute the most important part of the total costs of almost all chemical products [3.18]. The real production costs are derived from the retrospective calculation of production costs after the products have been made. However, for planning purposes the expected production costs must be known before production begins. They are obtained by cost estimates based on process-specific consumption data and taking market prices or specific transfer prices as valuation criteria.

The cost estimation of production costs for C_4-products of the plastics and motor fuel sectors in Chapter 6 of this book have been based on a standardised plan of classification of the types of cost. They contain information about the costs of raw materials, energy, labour and maintenance, and overhead and capital costs and even subtractable credits for co-products.

Raw material costs and credit items are derived from the engineering chemical consumption coefficients of raw materials (see Tables 4.4 for butane secondary products, 4.7 for butene secondary products and 4.11 for butadiene secondary products) in t raw material/t product (see Sect. 4.1) and of co-products in t co-product/t product, with a negative sign; and from the material-specific evaluations in DM/t raw material or co-product. These evaluations are based on market prices (published in the journals Europ. Chem. News and Chem. Marketing Reporter as spot market or contract prices) or transfer prices. The C_4-starting materials or co-products are evaluated in the processes of C_4-chemistry in this book as follows:

LPG, n-butane, refinery B-B (catcracker C_4-cut) at the heating value; higher prices than these do exist but in the FRG these are surplus co-products of refineries where the inferior use as refinery fuel determines the transfer price. An alternative value has been worked out for refinery B-B (in Sect. 6.1.3.2) based on the intended use as a gasoline component. However, this was not used in the cost estimations.

Butadiene-containing C_4-cut according to an estimate via the profit contribution of the contents of butadiene and raffinate I, minus the processing costs for butadiene extraction. The prices for this are negotiated mainly in long term contracts and are not published.

Raffinate I according to the own estimate of settlement price, since, as an unpriced co-product it is handled in the internal or external vertical production complexes. The transfer price is accordingly derived from a backwards calculation of its substitution value if it is used instead of refinery B-B for preparing MTBE and is due to yield MTBE production costs (see Table 6.8).

Raffinate II according to the own derived settlement price because, like raffinate I, it is an unpriced co-product. The evaluation should be orientated according to 1-butene, its most valuable constituent. The accounting price for raffinate II is derived retrospectively from a comparison of the cheapest acquisition of 1-butene from raffinate II (distillation) with the cheapest alternative preparation of 1-butene from a different base chemical which depends on the dimerisation of ethylene (IFP Alphabutol process) (see Table 6.25).

The evaluation of other materials is given in Table 5.3.

Energy costs are subdivided into costs for electricity, steam, fuel gas, refrigeration energy etc. The heating price valid for W. Europe in the middle of 1987 was established on the basis of the calculation below:

If firing is undertaken with boiler coal of thermal capacity $H_u = 29,000$ kJ/kg (corresponding to 7,000 kcal/kg = 1 ton coal equivalent, tce) at 110 $/t (average for indigenous and imported coal), a heating price of about 4 $/GJ (equivalent to 17 $/Gcal) emerges. The prices for large purchases of natural gas are of the same order of magnitude. The heating value of other hydrocarbons can be determined as evaluation ratios with this heating price via their calorific values.

The energy prices used in the tables of Chap. 6 are given in Table 5.4.

Labour and overhead costs are calculated as follows:

The manpower requirements had to be estimated when there were no literature data. A sum of 30,000 $/man-year is fixed for the costs of work of labour, 45,000 $/man-year for plant managers. The total personnel requirement is estimated to be five times the number of staff present in one shift.

Table 5.3. Compilation of the price data of the materials used in the cost estimation tables in Chap. 6 (mid 1987; DM/$ exchange rate 1.80)

Acetic acid	500 $/t
Acetone	410 $/t
Benzene	162 $/t (HV), 381 $/t (OV)
i-Butane	224 $/t
n-Butane	183 $/t (HV), 198 $/t (OV)
DMF	4 $/kg
Ethylene	410 $/t
FCC C_4-cut	183 $/t (HV)
H_2 (90%)	479 $/t
HF (98%)	1500 $/t
H_2SO_4 (98%)	100 $/t
Hydrogen cyanide	400 $/t
Methanol	130 $/t
NaOH (98%)	300 $/t
Nitrogen	35 $/1000 m_n^3
NMP	3 $/kg
Oxygen	40 $/t
Process water	1,50 $/m^3
Propane	186 $/t (HV)
Propene	380 $/t
Propene oxide	400 $/t
Raffinate I	280 $/t
Raffinate II	310 $/t
Steam cracker C_4-cut	400 $/t

HV: Heating value; OV: Octane value

Table 5.4. Compilation of the prices of the types of energy used in the cost estimation tables of Chapter 6 (mid 1987)

Boiler feed water	1.50 $ /m^3
Condensate	1.50 $ /m^3
Refrigeration energy	20 $ /GJ
Cooling water	7 ¢/m^3
Electricity	7 ¢/kWh
Heating gas	4 $/GJ (HV)
Steam	HP 16 $/t; MP 14 $/t; LP 12 $/t

HV: Heating value; HP: High pressure; MP: Medium pressure; LP: Low pressure

The overhead or indirect costs include general plant costs and are allowed for with an addition of 100% to the labour costs.

Capital costs are divided into those depending on fixed, and those on working capital, based on the necessary fixed and working capital requirements.

The time adjusted value for fixed assets (battery limits) has to be derived by projection from indices, provided the literature contains older data (Fig. 5.8). The Chemical Engineering Plant Cost Index gives information on data in American publications. German sources are covered by the Kölbel-Schulze (KS)-Preis-Index für Chemieanlagen. The KS-Index is published regularly in the periodical Chemische Industrie. Its average value of price increase in the period 1950–1986 was 4.73% p. a. [5.19]. Information in other currencies is converted via the rate of exchange at the particular time quoted in the publication. If there are no literature data about the requirement for fixed assets, other methods of estimation can be utilized [5.30]. Extra charges for offsites have not been taken into consideration. The costs related to invested capital cover depreciation and interest on the investments, based on an accepted 15 year period of economic use. An interest rate of 9% on the capital yields an annual repayment of about 12% (12.41% exactly). Estimates of 2% are made of the invested capital for taxes and insurances and 4% for service and maintenance. From this a total of 18% is derived for the costs related to the invested capital.

The requirement for working capital is taken as a uniform rate of 15% of the fixed capital. The costs depending on the working capital are taken into consideration as 9% p. a. of the working capital assets in order to cover the interest on capital. The capital requirement related to the unit of capacity, expressed in $/year-ton, is based on a plant utilization of 8000 hours per year.

The conditions in Western Europe serve as the basis for the cost estimations. These calculations must be suitably adapted when cost and price data in other countries deviate more markedly from these figures, e. g. through increased capital expenses in the developing countries. When price data from the area of validity of the FRG were used, the conversion was carried out at a rate of exchange of 1.80 DM/$.

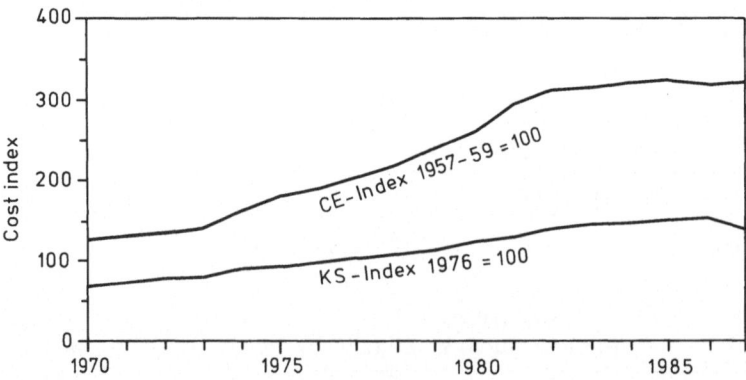

Fig. 5.8.
Chemical Engineering Plant Cost Index and Kölbel-Schulze-Preis-Index für Chemieanlagen

5.3.4.2 Forecasting Costs

Statements about the past and status quo are less interesting in the comparisons of production costs and full costs of competing products than are data showing how the differences in these costs are going to change in the future. If the rival products are made in processes in which the individual types of cost are differently structured, the relation of the production or total costs will unavoidably change with time because the individual types of cost will increase differently. An accurate forecast would require an,estimate of the influence of time on every single stage of the cost calculation over a period of time extending beyond that of the prognosis. This can be simplified by limiting the comparison to the four most important cost groups (raw materials, energy and labour costs, working and fixed capital related costs) and carrying out projection via the anticipated development of the cost indices. Electronic data processing is useful in this simulation.

5.3.4.3 Forecasting the Resource Costs of the Products

Forecasts of costs can be more accurate when based on the requirement of resources rather than on the costs, and then converting into costs afterwards. Energy costs thus become requirements in kWh, GJ, etc.; labour costs become the requirement of working hours of specialists and employees; material costs become the amounts needed of base chemicals and primary products per ton of product. Forecasts of the energy and material prices etc. must, however, be made. Extrapolation of the KS-price index yields sufficiently accurate results for the fixed capital costs (see Sect. 5.3.4.1).

5.3.4.4 Forecasts of Costs on the Basis of Energy and Crude Oil Equivalents

The raw material requirements and those of energy can be converted into, and expressed as, combined energy equivalents (MJ/kg, kWh/kg, kWh/l) for chemicals produced from petrochemicals and for the production of which the materials and energy are derived essentially from fossil sources. This greatly simplifies forecasting costs of production which depend closely on price increases of crude oil or natural gas. Possible changes in capital and labour costs are less influential in most large-scale technical processes, so that cost forecasting can be restricted to the effects of increases in the energy equivalents.

   This method of evaluation was developed during the oil crises because of a desire to be able to compare the raw material and energy contents of chemical and non-chemical products. The whole production hierarchy, from the geological deposits of the raw materials right up to the manufactured product was covered in many technical articles. Only the energy requirements of the erection and maintenance of the production installations were mostly not taken into consideration. The work of Kindler and Nikles, for example, quotes energy equivalents for most plastics and their base chemicals and intermediates [5.21, 5.22].

   The energy and raw material requirements can be expressed also as crude oil equivalents in a modification of this method. This has been performed in Sect. 6.1.6.2 for the example of motor fuel components.

6 Analyses of Competition of C₄-Chemical Products

6.1 Prospects for C₄-Technologies in the Field of Gasoline Components for Increasing Octane Numbers

6.1.1 The Need for High-Octane Components

The efficiency of conversion into propulsion performance of the energy content of fuel for internal combustion engines increases with a higher compression ratio of the fuel-air mixture on ignition. The theoretical relationship for Otto engines is given in Fig. 6.1.

If the compression of a gasoline-air mixture is too high, ignition "pockets" which tend to ignite spontaneously are formed in the cylinder. This is known as "knocking" of the engine, which reduces appreciably its degree of thermal efficiency and its service life. The anti-knocking ability of the gasoline is expressed in octane numbers and can be increased by adding special compounds. Octane numbers (ON) were introduced in 1927 as a measure of knocking resistance of fuels. They are measured by using test engines with variably adjustable compression ratios (in the FRG according to DIN 51756). The compression of the tested mixture is raised until the engine just begins to knock. A reference measurement is then carried out at the same compression ratio with mixtures of isooctane (2,2,4-trimethylpentane) of ON = 100 and n-heptane (ON = 0) until the engine just begins to knock again. The volume

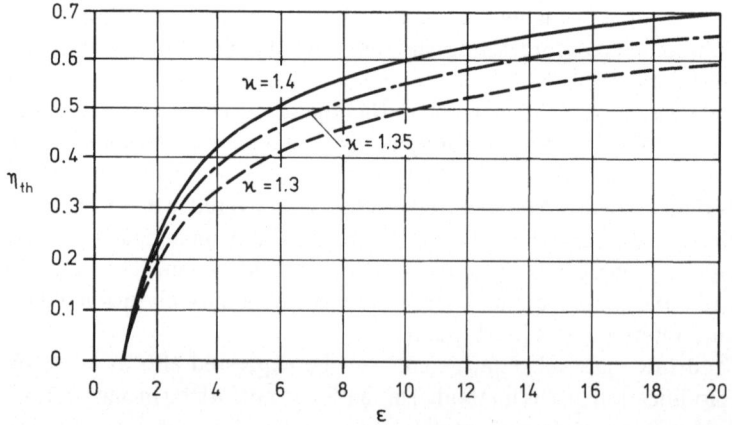

Fig. 6.1. Theoretical degrees of thermal efficiency of the Otto engine, as a function of the degree of compression (ε) and of the adiabatic exponent ($\varkappa$) [6.1]

proportion of isooctane in the mixture then gives the octane number of the tested mixture.

Depending on the conditions of the test procedure, distinction is made between:

RON	research octane number	test engine, rotational speed 600 r. p. m.
MON	motor octane number	test engine, rotational speed 900 r. p. m., mixture pre-heated to 150 °C
FON	front octane number	RON of the fraction of the fuel with B. P. up to 100 °C
	road octane number	RON measured with serial production motors on the road, or on the test stand

Nowadays it is the MON which rouses most interest, whereas it was formerly the RON.

Since gasoline fuels suddenly became more expensive in the 1970s success has been achieved with improving the efficiency of the Otto engine by increasing the permitted compression ratio of the combustion mixture; this was accomplished by:

1) Further development of the engines (design of the combustion chamber, electronic control of fuel supply and ignition).
2) Further development of Otto fuels (improvement of combustion properties under high compression, use of anti-knock components).

The following pages are devoted mainly to the successful contributions of chemistry and motor fuel refineries to the further development of fuels for the Otto engine.

It was customary up to 1983 to improve the octane number by adding lead alkyls because of their especially favourable price-performance ratio. Environment protection requirements have led to progressive reduction of the maximum permissible content of lead: in the FRG to 0.4 g/l in 1972 and to 0.15 g/l in 1976. The aim is, in connection with the introduction of exhaust gas catalytic converters, to replace leaded gasoline completely by lead-free fuel. Table 6.1 shows the legal requirements concerning lead content in the countries of Western Europe.

Table 6.1. Present legally fixed maximum contents of lead in gasoline (in g/l) in Western Europe, 1987 [6.2]

Austria	0.15; 0.013 (unleaded since Apr. 1985)
Belgium	0.40
Denmark	0.15
Finland	0.40
France	0.40
FRG	0.15 (leaded); 0.013 (unleaded)
Greece	0.40 (0.15 in Athens)
Ireland	0.40 since Jan. 1986
Italy	0.40
Luxembourg	0.40
Netherlands	0.15 since Oct. 1986
Norway	0.15
Portugal	0.40 since 1986
Sweden	0.15
Switzerland	0.15 (leaded); 0.013 (unleaded)
Spain	0.40 since 1986
United Kingdom	0.15 since Jan. 1986

The so-called gasoline pool made up of super (premium) and normal (regular) gasolines in the FRG had values of 96.0 for the RON and 86.0 for the MON in 1983 before unleaded gasoline was introduced. 0.15 g/l of lead alkyls was estimated to give an addition of 2.6 to the RON and 3.3 to the MON [6.3].

These deficiencies in the octane number due to replacement of lead-containing by lead-free gasoline, can be made up in two ways:

1) By further development work on the engines, aiming at higher compression ratios, even with gasolines of lower octane number, i. e. able to run without loss of thermal efficiency (reduction of the demand for high octane components).
2) By increasing the fuel quality using unleaded gasoline but of high-octane numbers (increase in supply of high-octane components).

Fulfilment of the former requirement would mean in principle that Otto engines with higher compression could use regular gasoline. The second requirement enables leaded premium gasoline to be replaced by unleaded.

These development goals would become superfluous if the diesel engine increasingly replaced the Otto engine in private cars. The diesel process operates at extremely high compression ratios of 22 to 24:1 and requires fuels of much less active combustion properties. In comparison to the spark ignition engine, which works mostly at compression ratios of 8 to 11:1, it has a much higher thermal efficiency. This results in a saving of fuel, which, together with partial elimination of the technical driving disadvantages of the diesel engine, have led to an increase in its share of the market as seen in the number of newly registered private cars in the FRG (Fig. 6.2). Further contributory factors are its acceptance as an environmentally positive engine because of its low carbon monoxide and hydrocarbon and more favourable nitrogen oxide emissions, the tax benefits, and, not least, the relatively low price per litre of diesel fuel together with its higher specific gravity giving it about a 13% higher volume-enthalpy content than the gasoline for Otto engines. Its share of the market

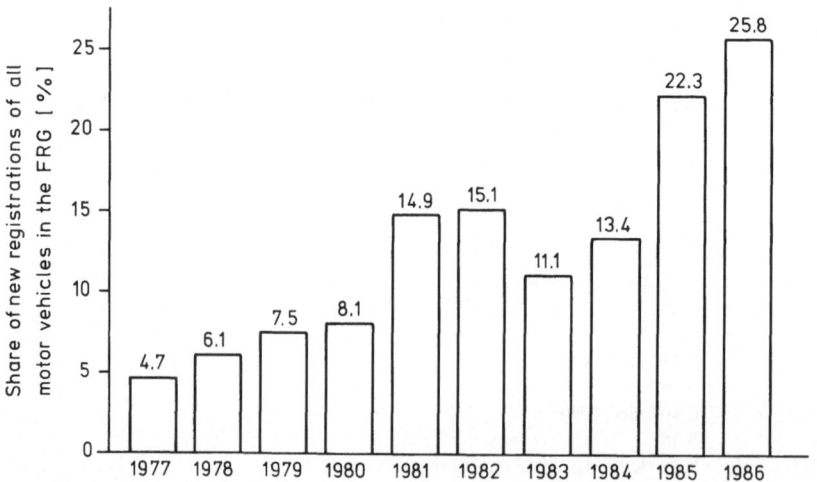

Fig. 6.2. Proportion of diesel vehicles among newly registered private cars and station wagons 1977–1986 [6.4]

could increase still further if a speed limit were introduced on motorways and the diesel engine were further developed.

The principal target for development is the so-called direct injection diesel engine for private cars, from which a decrease in consumption of up to 15% compared with that of the present generation of motors (pre-combustion chamber) is expected, provided that technical problems, such as noise, can be solved satisfactorily.

In contrast to these favourable forecasts of additional market success of the diesel engine there are other studies which indicate that the economical performance and the benevolence towards the environment have been over-estimated. According to the test results of a motor journal the diesel engine shows no advantage of lower gasoline consumption over the Otto engine when vehicle speeds exceed 120 km/h or 75 mph. Its superior performance is shown only in city traffic, with large yearly mileages and because of the distinctly lower diesel fuel prices [6.5].

The concentration of nitrogen oxides and the carbon particles in the exhaust gas which bind polycyclic aromatics create problems with diesel fuel. Catalysts and carbon filters would reduce still further the engine performance, which is not high in any case. This would lead to marked competitive disadvantages. At present toxicologists, government environmental experts and the Ministry for the Environment in the FRG are discussing the detrimental action of diesel exhaust gases for humans because long-term animal experiments have shown in the meanwhile that breathing these gases very probably causes lung cancer [6.6].

Figure 6.3 shows the stagnating sale of gasoline for Otto engines since the beginning of the 1980s in the FRG. Shell have forecast a fall in gasoline consumption in the private car sector to about 15.8 million tons by the year 2000 (see appendix to Fig. 6.3). Despite a clear increase in the number of diesel private cars (20% of the market) by the year 2000, ¾ of the fuel required for private cars will still be gasoline for Otto engines (21.1 million m^3) and ¼ diesel fuel (7.4 million m^3).

In recent years, about 20% of the FRG consumption of fuel for Otto engines has been imported, and about 10% of the annual gasoline production exported, principally to Switzerland and Austria. This yields a net import of just 10%, i.e., the amount of gasoline really produced in the FRG has been about 90% of the amount sold as shown in Fig. 6.3.

6.1.2 Material Balance in Optimising the Octane Number

Improvement of the anti-knocking property of gasoline leads to two opposing effects in the production and in the fuel use — an increase in octane number, i.e. an improvement of the gasoline quality, demands more energy and raw materials for the production, but the improvement in engine performance reduces the gasoline consumption per unit distance.

In order to solve this optimising problem ideas were developed to calculate the best octane number/material economy combination. The numerical values "Engine Response Factor" (ERF) and "Car Efficiency Parameter" (CEP) define the characteristic consumption feature, dependent on the engine, on varying the octane number. ERF is the numerical value of the Coordinating European Commission (CEC) for working out performance tests for lubricants and motor fuels. CEP is the

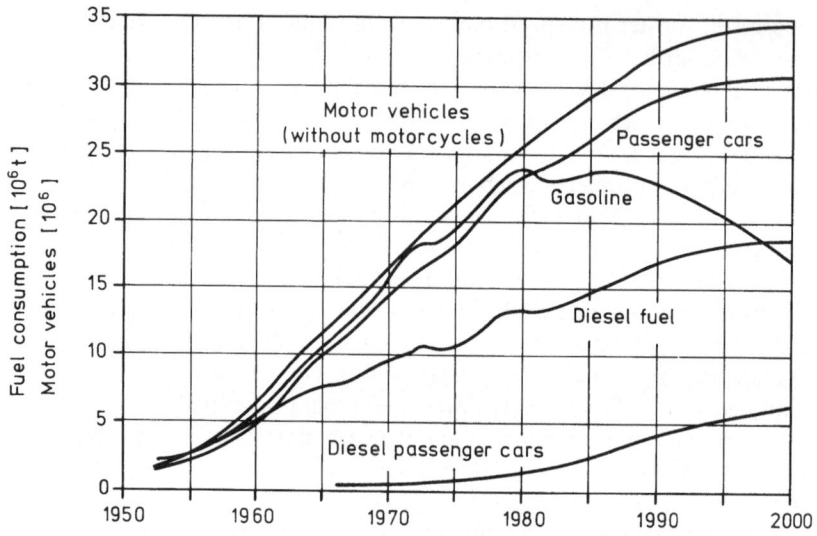

	1981			2000		
	Total	Gasoline	Diesel	Total	Gasoline	Diesel
Motor vehicles						
10^6	23.7	22.2	1.5	29.9	23.9	6.0
%	100	93.7	6.3	100	80.0	20.0
Fuel consumption						
10^6 t	23.5	20.7	2.8	22.0	15.8	6.2
t/car	0.99	0.93	1.87	0.736	0.661	1.03
m^3/car		1.24	2.23		0.881	1.237
km/a		11 925	25 200		10 800	19 500
l/100 km		10.4	8.9		8.2	6.3

Comparison 1981 / 2000 (%)

	Total	Gasoline	Diesel
Motor vehicles	+ 26	+ 7.7	+ 300
Consumption	− 7	− 24	+ 121
t/car	− 26	− 29	− 45
m^3/car		− 29	− 45
km/a		− 10	− 23
l/100 km		− 21	− 29

Fig. 6.3. Development of the number of motor vehicles and fuel sales in the FRG 1950–2000 [6.7]

corresponding value of the CONCAWE (Oil Companies' International Study Group for Conservation of Clean Air and Water in Europe). Studies showed that most popular Otto engines had CEP values between 0.7 and 1.2 (before catalysts were introduced), with a mean value of 1.0. This means that lowering the RON by one led to an increase of 1% by wt. in gasoline consumption [6.8].

A. D. Little, on behalf of the German Ministry for the Environment, as early as 1983 carried out a study of the effect on energy and raw material requirements in FRG refineries as a consequence of variations in octane quality [6.9].

If the higher or lower energy consumption in gasoline production and use are compared, this study yields the net increases in consumption of crude oil resulting from the change from leaded to unleaded gasoline shown in Fig. 6.4. The minimum of crude oil consumption is at the RON of 94.

Two interests conflict in this problem of the optimum: the efforts of car manufacturers to produce engines consuming the least possible amount of gasoline; and the attempts of refineries to produce gasoline qualities at the minimum energy costs. This optimum octane number is limited to optimising energy in the refinery and the gasoline requirements of the car. Consideration is given here to neither the larger costs of making engines with high compression ratios, nor the energy equivalents of calculable extra needs for preparing components for augmenting octane number.

Most car drivers prefer premium gasoline of octane numbers above this optimum value, so that engine knocking certainly does not occur even when driving with hard acceleration and at high speed. The complete replacement of a collective leaded gasoline of a quality corresponding to these customers' wishes (here simplified by

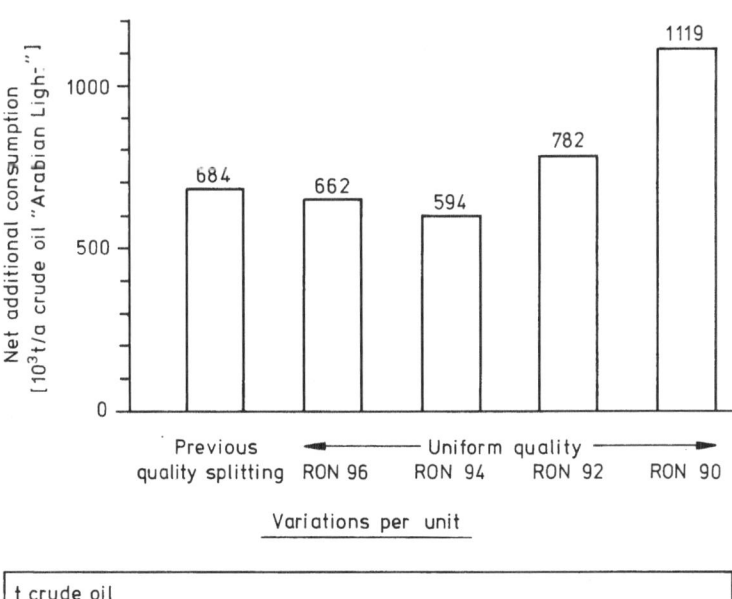

Fig. 6.4. Yearly net extra consumption, in crude oil units, for conversion of FRG refineries to production of 17.5 million t lead-free gasoline at CEP = 1 [6.10]

taking the quality of gasoline on sale in 1981*) would require an extra consumption of 684,000 t crude oil, based on a production of 17.5 million t gasoline (see Fig. 6.4). This statement is based on utilization at optimum cost of raw materials and processes with the costs and with the capacities of 1981.

The demand for octane improvers increased despite the stagnation in demand for motor fuel observed at the beginning of the 1980s (Fig. 6.3). This was the result of reduction in the supply of pyrolysis gasoline rich in aromatics. In addition, two of the suppliers, namely the naphtha crackers of Esso Chemie in Cologne and of Caltex in Raunheim, ceased production.

Catalytic reformate provides the principal part of the benzene also serving to improve octane number, and also alkylated aromatics for the gasoline pool. Benzene (RON = 108) is carcinogenic and should be kept out of the gasoline pool. The preparation of benzene in the FRG can be reduced by about 300,000 t/a by changing the operational procedures of the refineries, notably by feeding the catalytic reformers with higher-boiling fractions [6.11]. Lowering the benzene content still further to 0.6% by vol. would be extremely costly at the present state of benzene removal techniques (capital investment 1235×10^6 DM, running costs 1000 to 1100×10^6 DM/a [6.12]). What is to be done with the 500,000 t/a benzene remaining is not yet clear.

Catalytic cracking plants (mostly FCC plants) with their considerable contents of aromatics, are making an increasing contribution to meeting the demand for high-octane fuel. The capacity of the catalytic cracking plants in the FRG has remained constant at about 10 million t/a. On the other hand, the reformer capacity fell between 1983 and 1986 from 18.5 to 13.9 million t/a, mainly as a result of the closing down of refineries [6.13].

6.1.3 Technical Alternatives in Production of Gasoline of High Octane Number

6.1.3.1 Survey

The increasing demand for gasoline of higher octane number can be met in various ways. Decisive factors influencing the choice of suitable processes are: the availability of raw materials; the economic success of the production of the components which increase the octane number; and any synergistic effects arising in the gasoline pool. The economic success also depends on the refinery flexibility.

Table 6.2 contains the most important components of future motor fuels, i. e. without lead alkyls and benzene. It is evident that the great majority of them are based on the C$_4$-hydrocarbons. The choice of the most suitable components is governed by specific criteria. The most important criteria are volatility, density, and octane rating.

---

* octane qualities in 1981 [6.14]:

	RON	MON	splitting (vol%)
premium	98.6	88.2	54.0
regular	92.9	84.1	46.0

Table 6.2. Motor fuel components of high octane number and their base materials

Base materials	Process	Products
C$_4$-Hydrocarbons		
i-butane + butenes	alkylation	butene alkylate
i-butane + propene	alkylation	propene alkylate
i-butane + propene + O$_2$	Arco PO process	GTBA
i-butene	dimerisation	butene dimerisate
i-butene	oligomerisation	butene polymer gasoline
i-butene + H$_2$O	hydration	TBA
i-butene + methanol	MTBE process	MTBE
i-butene + ethanol	ETBE process	ETBE
n-butenes	hydration	SBA
Other Hydrocarbons		
propene	dimerisation	propene dimerisate
propene	oligomerisation	propene polymer gasoline
propene	hydration	IPA
propene	OXO-synthesis, hydrogenation	IBA
i-pentene + methanol	TAME process	TAME
n-pentane/*n*-hexane	isomerisation	*i*-pentane/*i*-hexane isomerate
naphtha	reforming	reformate
naphtha	steam cracking	pyrolysis gasoline
vacuum gas oil	FCC cracking	crack gasoline
crack gasoline	Etherol process	mixed ethers
synthesis gas	methanol synthesis	methanol
synthesis gas	fuel-methanol synthesis	methanol with higher alcohols
synthesis gas	Ensol process	ethanol
ethylene	ethanol synthesis	ethanol
biomass	fermentation	ethanol

6.1.3.2 Direct Blending of Refinery B-B

Increase of octane number with refinery butanes and butenes is especially convenient for the refineries because they can be directly blended with the fuel in a technically simple way, i. e. without loss of material and additional processing costs (Table 6.3). Their miscibility is restricted by the standardised maximum vapour pressure of the fuels (see Sect. 3.2.7).

The transfer price used for refinery B-B in the fuel calculation is at the same time a guide for their transfer price in the event of an alternative use in another refinery or

Table 6.3. Properties of fuels from refinery B-B

Property	refinery B-B	1-butene	2-butene	*i*-butene	*n*-butane	*i*-butane
RON	104	71	106	102	93	100
MON	81	79	99	99	92	96
Vapour pressure according to Reid (in bar)	3.9	3.5	3.8	5.0	3.6	5.3

chemical process. The calculation of this reference value is presented in a following example, taken from 1983 because at that time in Western Europe only the leaded qualities of premium and regular gasolines were in use and unleaded gasoline was not available.

The yardsticks for evaluation are, on the one hand, the market prices for gasoline (regular and premium) and aromatics, and, on the other hand, the octane numbers of the fuels. Production statistics [6.15] show, that in the second quarter of 1983 in the FRG, the statistically produced amount of 2,535 million t of gasoline corresponded to 1,032 billion (10^9) DM for the market production (i. e. proceeds before turnover tax). From this can be calculated an average value of the combined premium and regular gasoline produced for sale, of about 400 DM/t designated hereafter as production value. Taking a sales ratio of premium/regular of 53/47 and the price difference between them of 0.06 DM/l, corresponding to 80 DM/t based on a fuel density of 0.75 kg/l, the individual data for premium and regular gasoline can be calculated from the equation:

$$0.53(x + 80) + 0.47x = 400$$

where x is the price of regular gasoline in DM/t. Table 6.4 lists the production values in DM/t and $/t, and also the corresponding octane numbers of the motor fuels.

Table 6.4. Production and octane values of motor fuels in the FRG, 1983

	Production value		Octane number	
	DM/t	$/t	RON	MON
Regular gasoline	357	198	93.0	84.0
Premium gasoline	437	243	98.5	88.5

From this a transfer price for refinery B-B in the gasoline pool can be derived by relating the production values and the octane numbers. In Fig. 6.5 the data of Table 6.3 are plotted, taking the two points, for premium and regular gasoline. These are joined by a straight line as a simplifying approximation. Extrapolation and dropping perpendiculars from the refinery B-B octane values to the other axis yields the transfer prices corresponding to the production values. These are between 323 (RON) and 168 (MON) in $/t.

This model calculation is valid only under the following conditions:

— the fuel components behave like ideal mixtures, i. e. the octane number of the mixture, ON, is directly proportional to the octane number and to the mass fraction, m, of each component, according to the relationship:

$$ON = \sum_{i=1}^{n} ON_i \cdot m_i .$$

possible synergistic effects are not taken into consideration.
— there is a linear relation between the production value and the octane number.

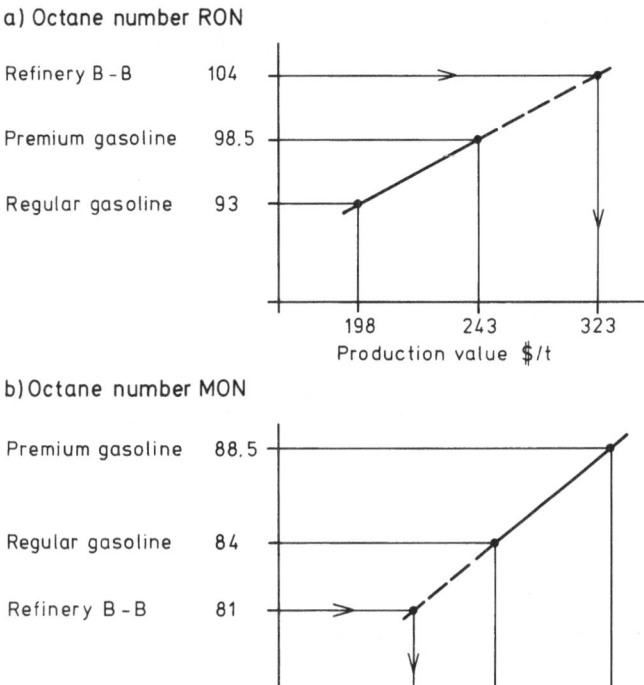

a) Octane number RON

b) Octane number MON

Fig. 6.5. Calculation of the transfer price for refinery B-B

The relatively low MON of the butene components, especially of 1-butene depresses the transfer price of refinery B-B. It would therefore be preferable to use only the butanes (refinery gas) to improve the octane number and to convert the butene moiety of refinery B-B into motor fuel components of higher value or into chemical products.

In practice, the vapour pressure limit restricts the added amount of refinery B-B to 4 to 6% by weight, so that the octane number can be only moderately influenced in this way. This, however, does not diminish the value of the model calculation performed above.

Fuel oil and heavy oil are relatively more in demand in the FRG than in the USA. Less crude or heavy oil has therefore been converted in the FRG for a long time. This explains the large proportion of catalytic cracking in the USA (Table 6.5).

In the USA especially large amounts of refinery B-B are produced and far more of it is converted into liquid components of high octane number, than in the FRG.

6.1.3.3 Conversion of Refinery B-B into Liquid Components

Numerous technologies are available for converting refinery B-B into liquid gasoline components. These have been treated above and it is sufficient here to name the relevant classes of products and quote the pertinent sections (see Table 6.6).

Table 6.5. Production and absorption of refinery B-B in motor fuel production. Comparison of the USA and the FRG, 1983

Line		USA		FRG	
1.	Crude oil throughput	ca. 800	MMt/a	ca. 70	MMt/a
2.	Production from catalytic crackers	250	MMt/a	9.4	MMt/a
3.	Yield of refinery B-B (9 wt. % of line 2)	22.5	MMt/a	0.846	MMt/a
4.	Amount of gasoline produced	320	MMt/a	20	MMt/a
5.	Max. absorption amount on refinery B-B (7 wt. %	22.44	MMt/a	1.4	MMt/a
6.	of line 4)				
7.	Preparation of alkylate gasoline	35	MMt/a	0.31	MMt/a
8.	Proportion of catalytic cracking (line 2/line 1)	0.31		0.13	
	Proportion of conversion of crude oil into gasoline (line 4/line 1)	0.40		0.28	

Table 6.6. References regarding the conversion of refinery B-B into liquid gasoline components

Component	Starting material	Section in this book
Dimerisate gasoline	propene and butenes	3.3.2.2
Polymer gasoline	propene and butenes	3.3.1.1
Alkylate gasoline	propene and butenes, with i-butane	3.2.4
TBA	i-butene	3.1.2.2;
	or propene with i-butane	3.1.2.4
SBA	n-butenes	3.1.2.2; 3.1.2.4
MTBE	i-butene	3.3.1.2

6.1.4 Comparison of Quantities

The attainment from crude oil of maximum yields of gasoline in quality suitable for sale is the prime target of most refineries. Simply equipped refineries, without units for subsequent processing of heavy fractions from the distillation plants, are able to influence the yield of gasoline only through their choice of crude oil type, from which, depending on its source, more or less light straight run gasoline can be distilled. Until the first oil crisis in 1974 these so-called hydroskimming refineries were the standard refinery type of Western Europe, especially the FRG [6.16]. In that form they have almost disappeared after the comprehensive modifications during the past 15 years. More complex refineries are able to augment the yield of gasoline by after-treatment (conversion), e. g. of gas oil (catalytic cracking) and/or residues, such as heavy fuel oil (thermal cracking).

A fuel of suitable marketable quality can be obtained from light gasoline prepared in this way, only by subjecting it to refining to increase its octane number. Most refineries accomplish this predominantly by reforming the gasoline fraction. Occasionally isomerisation is also applied. The octane number can be increased by isomerisation of the C_5/C_6-gasoline fraction but only at the expense of its output (see case 5 in Table 6.7). In the same way, more severe reforming conditions lead to considerable naphtha losses because gaseous hydrocarbons are formed.

Increase in output and improvement of octane number at the same time are attained when additional liquid components for ON improvement are made from excess C$_3$- or C$_4$-hydrocarbons (refinery gas, refinery B-B, steam cracker cuts).

A description has been given in the literature of an example of a medium-sized refinery (crude oil throughput of 100,000 barrels/day = 4.7 MMt/a Arabian Light) equipped with reforming, cracking, coking and alkylation units [6.17]. The output amount changed when:

— the refinery additionally converted propene into gasoline components,
— a prior stage was introduced so that the *i*-butene part of the FCC exit gas was not alkylated but used for producing MTBE,
— the improvement in octane number was achieved by blending purchased TBA/methanol to achieve the maximum legally permitted concentrations.

Table 6.7 shows the differences in output from a base case. Comparison of the procedures 1 to 5 shows that the amount of gasoline in case 1 (C$_3$-alkylation) was the most favourably influenced; next was case 4 (TBA/methanol blending). Case 2 (C$_3$-dimerisation or -polymerisation) yielded only a small improvement in output and case 3 (MTBE production) reduced the amount of gasoline.

Case 3 illustrates the limited possibilites of MTBE production in this relatively well equipped refinery. If the whole *i*-butene content of the crack gases of the FCC plant are diverted for producing MTBE, an extremely small amount of MTBE is obtained, contributing only 1.1% by vol. to the gasoline pool. Expressing the example in absolute figures, the gasoline production of 8.626 million l/day corresponds to an output of 2.3 MMt/a gasoline containing 25,300 t MTBE.

The amount of MTBE could be significantly increased by purchasing C$_4$-cut of ethylene steam crackers or by preparing it from the refinery gas butanes by isomerisation and dehydrogenation.

There is a special advantage in all the variants in blending the highest permissible amount of butanes. The butane fraction is particularly large in case 1, C$_3$-alkylation, which gives an increase of 12% over the basic case; and smallest in case 4, TBA/methanol blending, which gives a decrease of 45% butanes. MTBE scarcely alters the vapour pressure of the gasoline pool and thus has no negative influence on the butane addition.

Blending of MTBE (case 3) or TBA/methanol (case 4) increases the octane number so much that production of the most expensive component (butene alkylate) can be reduced by 15% in both cases. The average octane numbers (RON + MON)/2 of 87.5 or 88.8, due to the high ON values of the components, accordingly are clearly above that of the base case.

The arithmetical product of the daily gasoline output and the average octane number of the gasoline pool yielded is termed "octane-barrel" in the American oil refining industry and is a customary measure for the quantitative performance of complex refineries. The lowest line of Table 6.7 determines the litre-octane-performance, a term which takes into account the qualitative differences of the gasoline as far as the MON is concerned, and which is adjusted to European units, for characterising the refining processes (see Sect. 6.1.6.2).

Table 6.7. Comparison of the gasoline outputs of a minimum tolerable quality (RON = 95, suitable volatility) from various processes [6.17]

Constants	
crude oil throughput	15.9 × 10⁶ l/day equal to 100,000 bbl/day or 4.7 × 10⁶ t/a
straight run gasoline	1.43 × 10⁶ l/day
reformer throughput	2.75 × 10⁶ l/day reformer severity at RON = 95
cracker throughput	2.78 × 10⁶ l/day
sulfuric acid alkylation of all butenes yielded (cases 3 and 4 limited)	
residue coked	
propane utilized as refinery fuel	

Case	0	1	2	3	4	5
Manufacturing process	base case	C_3-alkylation	C_3-dimerisation or -polymerisation	MTBE	TBA/methanol	C_5/C_6-isomerisation
Gasoline components, vol %						
Butanes	9.2	9.7	9.2	9.2	4.8	8.7
Straight run	16.5	15.5	16.2	16.6	15.7	32.1
Reformate	31.7	29.9	31.0	31.8	30.2	32.5
FCC	32.1	30.2	31.4	32.3	30.8	10.6
Butene alkylate	10.5	9.9	10.3	9.0	8.5	
Propene alkylate		4.8				
Propene dimerisate or polymerisate			1.9			
MTBE				1.1		
TBA/methanol (50/50)					10.0	
C_5/C_6-isomerate						16.1
Gasoline output 10⁶ l/day	8.666	9.190	8.856	8.626	9.079	8.576
MON	79.6	80.0	79.6	80.0	82.6	83.4
Litre-octane-performance 10⁶ l × MON/day	689.8	735.2	704.9	690.1	749.9	715.2

to case 1: additional feed of the highest possible obtainable amount of 246,500 l/day propene from the refinery gas for the sulfuric acid alkylation

to case 2: IFP Dimersol process or catalytic UOP polymerisation for 246,500 l/day propene

to case 3: MTBE process arranged in front of the alkylation, and conversion of the maximum amount of i-butene co-product from the FCC process

to case 4: maximum permitted blending according to EPA (3.5% by wt. of Oxygen in the gasoline pool, corresponding to a mixture with 10% by vol. of 50/50 TBA/methanol) of a purchased component (e.g. Oxinol 50)

to case 5: installation of a C_5/C_6-reflux isomerisation for throughput of all the straight run gasoline

6.1.5 Comparison of Costs

Alkylation, polymerisation, production of MTBE, TBA, and SBA are the most important processes for obtaining gasoline components from C$_4$-hydrocarbons.

The cost comparison is carried out on the basis of production costs per ton for the respective motor fuel components. Reference is made to Sect. 5.3.4.1 for information about the assumptions underlying the determination of these production costs.

6.1.5.1 Production Costs for Alkylate Gasoline

The raw materials for producing alkylate gasoline by the Stratco sulfuric acid and hydrofluoric acid processes are obtained from the exit gas stream from catalytic crackers (FCC-cut for the HF process freed from sulfur and dried). It must be enriched with *i*-butane.

The calculation (Table 6.8) refers to a mixture containing only little propene (5% by vol.). This keeps low the consumptions of sulfuric acid in the process using it, and of *i*-butane in the HF process. Propene alkylate requires three times as much sulfuric acid as butene alkylate. The acid consumption in the HF process is, on the other hand, independent of the olefine used. In contrast, increase in the specific consumption of *i*-butane in the presence of propene is observed only in the HF process. This is explained by the so-called hydrogen transfer when *i*-butane is in excess:

1 mole propene + 2 moles *i*-butane → 1 mole propane + 1 mole C$_8$-alkylate.

As a result of the larger acid requirement, the catalyst costs of the H$_2$SO$_4$ process are ten times those of the HF process. The calculation is based on a capacity of 124,000 t/a alkylate. Such a capacity delivers about 10,000 t/a of dilute sulfuric acid, contaminated with organic materials. This amount is difficult to sell and difficult to dispose of without environmental problems. On the other hand re-concentration of the acid within the factory is costly.

Most H$_2$SO$_4$-alkylation plants in the USA are situated near the sulfuric acid producers so that the waste acid can be reprocessed with minimum expense. This reconcentration using external equipment is also included in the calculation and estimated as 100 $/t sulfuric acid.

In the HF process an acid side stream, which can be simply separated from the alkylate, is fed to a regeneration column. The low losses of HF occur mainly in the bottom product of this regeneration step where acid tar and an HF/H$_2$O-azeotrope are taken off. Traces of acid in the product stream after the purifying distillation of the alkylate have to be washed out; for this either NaOH (H$_2$SO$_4$ process) or KOH solid bed filtration (HF process) are used, which add negligible costs.

The total energy costs of the two processes scarcely differ but the structures of the costs do, leading to deviating data because of differing transfer prices for steam and heating gas. The separation of the product in the H$_2$SO$_4$ process is carried out in a number of steam-heated columns, whereas in the HF process, circulated alkylate is superheated in an oven operated with heating gas thereby providing the heat energy for thermal defluorination and further product separation.

It must be borne in mind in the alkylation processes that the raw material must always be enriched by small amounts of additional *i*-butane. About 8−10% by wt. of

Table 6.8. Cost of production for C$_6$-alkylate from Stratco processes with H$_2$SO$_4$ and HF

Product, purity:	C$_{6+}$-Alkylate, 88.5 wt.%			C$_{6+}$-Alkylate, 88.2 wt.%		
Raw material, specification:	FCC C$_4$-Cut			FCC C$_4$-Cut		
Process:	Stratco H$_2$SO$_4$-Alkylation			Stratco HF-Alkylation		
Capacity:	140,000 t/a Alkylate			140,000 t/a Alkylate		
Capital investment	10^6 $	10^6 $	$/t · a	10^6 $	10^6 $	$/t · a
Basis	(III 85)	(III 87)		(III 85)	(III 87)	
Fixed capital	14.9	14.7	105	14.5	14.3	102
Working capital		2.2	16		2.1	15
Total capital investment		16.9	121		16.4	117

Production costs	Unit costs	Units/t	Costs $/t		Units/t	Costs $/t
Raw material costs						
FCC C$_4$-Cut 45.7 GJ/t	4 $/GJ	1.31 t	239		1.32 t	241
(34 vol% Butenes, 5 vol% Propene)						
i-Butane	224 $/t	0.08 t	18		0.10 t	22
	(see					
	Table 6.12)					
Catalyst and H$_2$SO$_4$ (98%)	100 $/t	0.08 t	8			
auxiliaries costs HF (98%)	1.5 $/kg				0.5 kg	1
Energy costs						
Electricity	7 ¢/kWh	92 kWh	6		26 kWh	2
Steam HP 42 bar	16 $/t	0.41 t	7			
MP 10 bar	14 $/t	0.15 t	2		0.14 t	2
LP 4 bar	12 $/t	0.12 t	1			
Cooling water	7 ¢/m^3	58 m^3	4		72 m^3	5
Heating gas	4 $/GJ				2.55 GJ	10
Material and energy costs			285			283
Labour and overhead costs						
Operating labour	30000 $/my	3/shift	3		3/shift	3
Operating supervision	45000 $/my	1	0		1	0
Plant overhead	100% excess		3			3
Capital related costs						
Fixed capital related costs	18%		19			18
Interest on working capital	9%		1			1
By-product credit						
Propane 46.4 GJ/t	4 $/GJ	−0.06 t	−11		0.07 t	−13
Total cost of production $/t			300			295

Source: [6.18; 6.19]

i-butane is required, referred to the yielded amount of alkylate. At the capacity of the example, this means about 10,000–12,000 t/a *i*-butane. This amount is too small to make butane isomerisation economical. Butamer plants licensed by UOP produce from 40,000 to 680,000 t/a *i*-butane [6.20]. In this case it was assumed that an *i*-butane joint production exists and that small amounts from that plant can be supplied. The production costs of an isomerisation plant with a capacity of 351,000 t/a *i*-butane was calculated in Table 6.12. The production costs have been introduced.

The cost estimation (Table 6.8) shows that the production costs for butene alkylate are 300 $/t for the H$_2SO_4$ process and 295 $/t for the HF process. The costs of disposal of the acid tar unavoidably formed (5 kg/t alkylate in the H$_2$SO$_4$ process and 2 kg/t in the HF process) depend considerably on the local conditions.

6.1.5.2 Production Costs for Polymer Gasoline

Table 6.9 contains a list of the estimated production costs of polymer gasoline by the Selectopol process of IFP. The raw material used here, raffinate I, presupposes the presence of a steam cracker (raw materials: gas oil, naphtha or LPG but not ethane) on the refinery site or in the immediate neighbourhood.

Only a little C$_4$-cut is yielded in the USA where steam crackers are predominantly fed with ethane. Refinery B-B is mainly used there for producing polymer gasoline.

The catalyst costs in the IFP process can be kept low by using a secret, especially

Table 6.9. Cost of production for polymer gasoline by the IFP process

Product, purity:	Polymer gasoline		
Raw material, specification:	Raffinate I, 48 wt. % i-Butene		
Process:	IFP Selectopol		
Capacity:	50,000 t/a Polymer gasoline		
Capital investment	10^6 $	10^6 $	$/t · a
Basis	(II 80)	(III 87)	
Fixed capital	1.6	2.0	40
Working capital		0.3	6
Total capital investment		2.3	46
Production costs	Unit costs	Units/t	Costs $/t
Raw material costs			
Raffinate I	280 $/t	214 t	599
	(see Table 6.10)		
Catalyst and auxiliaries costs			2
Energy costs			
Electricity	7 ¢/kWh	21 kWh	2
Steam MP 15 bar	14 $/t	0.35 t	5
LP 5 bar	12 $/t	0.19 t	2
Cooling water	7 ¢/m^3	40 m^3	3
Material and energy costs			613
Labour and overhead costs			
Operating labour	30000 $/my	1/shift	3
Operating supervision	45000 $/my		
Plant overhead	100% exc.		3
Capital related costs			
Fixed capital related costs	18%		7
Interest on working capital	9%		1
By-product credits			
Raffinate II	310 $/t	−1.12 t	−347
	(see Table 6.24)		
Total cost of production $/t			280

Source: [6.21]

stable carrier catalyst, probably similar to that of the UOP process (phosphoric acid on kieselguhr). The process gives a very pure, butene-containing raffinate II which, under favourable conditions (taken as being present here), yields greater credit items than a refinery gas evaluated only according to its heat equivalent.

The production costs of the polymer gasoline according to the IFP process amount to 280 $/t. They would be of the same order of magnitude if refinery B-B were used as raw material.

6.1.5.3 Production Costs for MTBE

The customary processes for making MTBE use the C₄-cut from steam crackers or catalytic crackers. How these alternatives affect the production costs is explained here using the examples of the Hüls MTBE process which employs steam cracker C₄-cut, and that of Neochem which uses FCC C₄-cut (see Table 6.8).

The FCC C₄-cut fed into the Neochem plant contains less i-butene than the raffinate I used in the Hüls process. About three times as much feedstock must thus be used at the same capacity (100,000 t/a). If the amount of co-products (see by-product credit) is taken into consideration, there is no net additional consumption of C₄-cut in the case of the FCC stream. The plant must, however, be larger, which increases the fixed capital investment by about one-third. The specific energy consumption on using the FCC cut is also increased, by a factor of three. This is mainly true because the specific steam consumption is nearly inversely proportional to the i-butene concentration in the raw material (see Fig. 6.6).

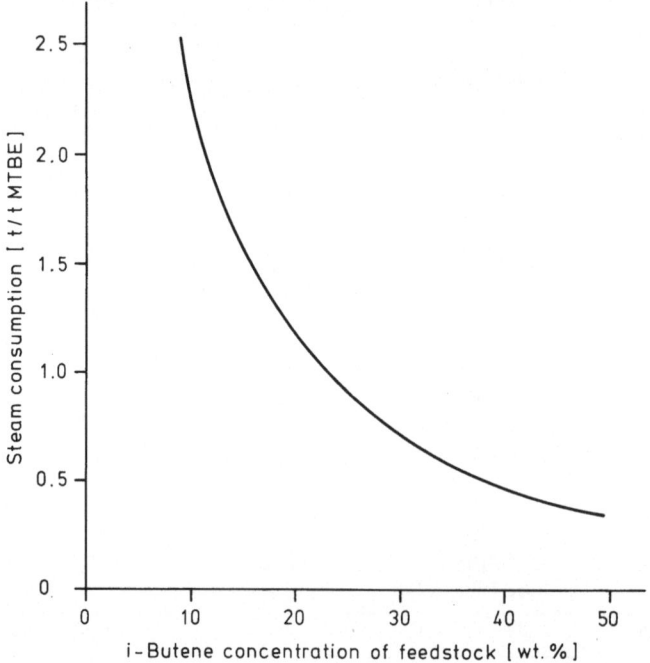

Fig. 6.6. Specific steam consumption of MTBE production [6.24]

The production costs for MTBE from FCC C$_4$-cut (refinery B-B) are 218 \$/t. The most important raw material competition for MTBE production is from alternate use of raffinate I or refinery B-B. If the production costs for MTBE in the Hüls process from raffinate I are set at the equal amount of 218 \$/t, a retrograde calculated transfer price of raffinate I of 280 \$/t is obtained.

Table 6.10. Cost of production for MTBE by processes from Hüls and Neochem

Product, purity:	MTBE, 99 wt.%			MTBE, 95 wt.%		
Raw material, specification:	Raffinate I, 44 wt.% *i*-Butene			FCC C$_4$-Cut, 18 wt.% *i*-Butene		
Process:	Hüls MTBE			Neochem MTBE		
Capacity:	100,000 t/a MTBE			100,000 t/a MTBE		
Capital investment	10^6 \$	10^6 \$	\$/t · a	10^6 \$	10^6 \$	\$/t · a
Basis	(II 80)	(III 87)		(II 80)	(III 87)	
Fixed capital	3.1	3.5	35	4.2	5.2	52
Working capital		0.5	5		0.8	8
Total capital investment		4.0	40		6.0	60
Production costs	Unit costs	Units/t	Costs \$/t		Units/t	Costs \$/t
Raw material costs						
Raffinate I	280 \$/t	1.5 t	420			
FCC C$_4$-Cut 45.7 GJ/t	4 \$/GJ				4.1 t	749
Methanol	130 \$/t	0.37 t	48		0.38 t	49
Catalyst and auxiliaries costs			0			2
Energy costs						
Electricity	7 ¢/kWh	5 kWh	0		15 kWh	1
Steam MP 15 bar	14 \$/t	0.4 t	6		1.3 t	18
Cooling water	7 ¢/m^3	26 m^3	2		80 m^3	6
Material and energy costs			476			825
Labour and overhead costs						
Operating labour	30000 \$/my	0.5/shift	1		0.5/shift	1
Operating supervision	45000 \$/my					
Plant overhead	100% excess		1			1
Capital related costs						
Fixed capital related costs	18%		6			9
Interest on working capital	9%		1			1
By-product credit						
Raffinate II	310 \$/t	−0.86 t	−267			
LPG 45.7 GJ/t	4 \$/GJ				−3.39 t	−619
Total cost of production \$/t			218			218

Source: [6.22; 6.23]

6.1.5.4 Production Costs for TBA

The raw materials raffinate I or *i*-butane/propene serve for producing TBA. The latter route dominates. The production costs of the Hüls TBA process and those of the Arco propene oxide process (with TBA as co-product) are compared in Table 6.11.

Table 6.11. Cost of production for TBA by processes from Hüls and Arco

Product, purity:	TBA, 91 wt.%			TBA, 95 wt.%		
Raw material, specification:	Raffinate I, 39 wt.% *i*-Butene			*i*-Butane, Propylene		
Process:	Hüls TBA			Arco Propene oxide with TBA		
Capacity:	350,000 t/a TBA			150,000 t/a PO with 380,000 t/a TBA		
Capital investment	10^6 $	10^6 $	$/t · a	10^6 $	10^6 $	$/t · a PO + 2.5 t TBA
Basis	(III 82)	(III 87)		(I 86)	(III 87)	
Fixed capital	11.3	11.6	33	300	300	566
Working capital		1.7	5		45	85
Total capital investment		13.3	38		345	651
Production costs	Unit costs	Units/t	Costs $/t		Units/t	Costs $/t
Raw material costs						
Raffinate I	280 $/t	2.20 t	616			
i-Butane (see Table 6.12)	224 $/t				2.25 t	504
Propene	380 $/t				0.80 t	304
Oxygen	40 $/t				1.25 t	50
Process water	1.50 $/m^3	0.31 t	1		1.95 t	3
Catalyst and			1			
auxiliaries costs Mo	30 $/kg				0.6 kg	18
Energy costs						
Electricity	7 ¢/kWh	19 kWh	1		400 kWh	28
Steam MP 10 bar	14 $/t	0.56 t	8		5 t	70
Cooling water	7 ¢/m^3	24 m^3	2		n. a.	–
Material and energy costs			629			977
Labour and overhead costs						
Operating labour	30000 $/my	3/shift	1		20/shift	6
Operating supervision	45000 $/my	1	0		3	0
Plant overhead	100% excess		1			6
Capital related costs						
Fixed capital related costs	18%		6			102
Interest on working capital	9%		1			8
By-product credit						
Raffinate II	310 $/t	−1.51 t	−468			
LPG 45.7 GJ/t	4 $/GJ				−0.08 t	−15
Cost of production $/t PO + 2.5 t TBA						1084
% Market price $/t PO 700 $/t less $ 70, net 630 $/t						−630
$/2.5 t TBA						454
Total cost of production $/t TBA			170			182

Source: [6.25; 6.26]

The cost estimation deviates from the usual methods in the case of the Arco process. This is because the production costs are first determined in relation to the total output of the two co-products ($ per t propene oxide plus 2.5 t TBA). The propene oxide is more valuable but the yield is smaller than the TBA. If propene oxide is regarded as a by-product and its market value is subtracted from the production costs of the whole co-production, the net costs of TBA may be found.

Sales and administrative expenses, however, have to be deducted from the market value in order to obtain the net credit for the propene oxide. The sales and administrative costs are estimated to be 10% or 70 $/t propene oxide.

The results are production costs for TBA from raffinate I of 170 $/t and from *i*-butane/propene as 182 $/t. TBA is always blended with methanol as a fuel component, so that the price of the mixture is always influenced by the methanol price. It is usual to mix in a 50:50 ratio, as in Oxinol 50 of Arco, for instance.

Only the production costs of a crude TBA containing about 9% water is considered in the calculation for the TBA process of Hüls. This must be dehydrated to about 99% TBA before use as a fuel component. This increases the production costs by a small amount. The Arco TBA contains about 5% impurities (including about 1% water by wt.) but these hardly interfere in its use as a fuel component.

Figure 6.7 contains production costs for TBA according to the Arco process, expressed in relation to various raw material prices and net credits for the co-product propene oxide.

6.1.5.5 Production Costs for MTBE and TBA from Butanes

The production of MTBE and TBA according to the principal processes cannot be increased at will. MTBE production can be raised only in dependence on the amounts of *i*-butene-containing C$_4$-cuts obtained as by-product from steam crackers and

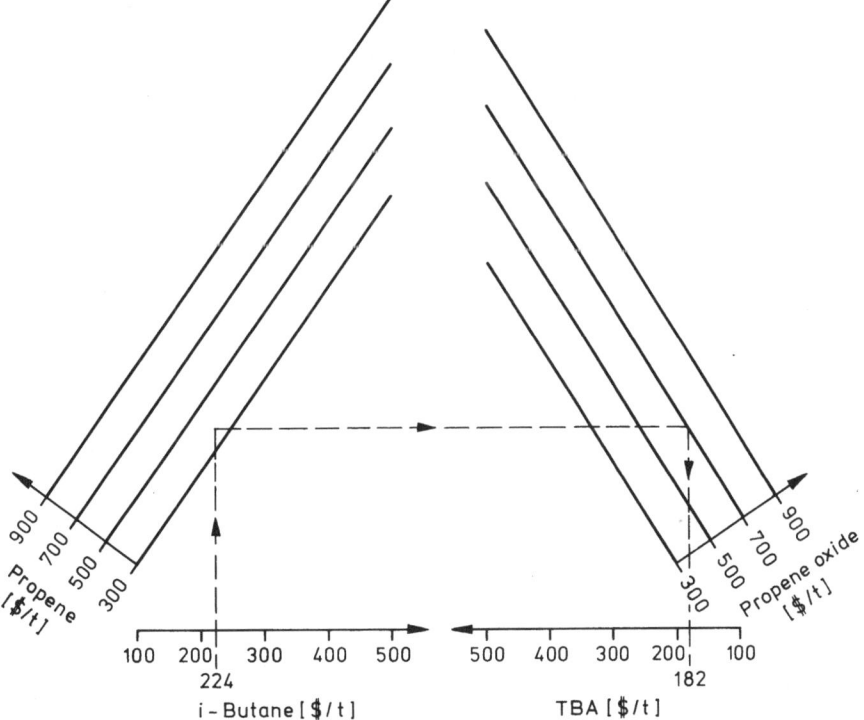

Fig. 6.7. Production costs of TBA in relation to prices of raw materials and market prices of co-product, according to the Arco process

refineries. The production of TBA can likewise only be expanded according to the market potential of the co-product, propene oxide, of the Arco process. The supply limit, already mentioned, applies also to TBA production on the basis of i-butene-containing C_4-cuts from cracking. As will be shown below (Sect. 6.1.7) complete replacement of the lead alkyls would induce a high demand for MTBE and TBA. This could lead to a widening of the raw material basis, including the butanes.

It would be especially interesting here to use the butane contained in natural gas and particularly that associated with petroleum reserves in combination with the methanol which has hitherto been produced in large amounts via synthesis gas in the

Table 6.12. Cost of production for i-butene from n-butane via isomerization/dehydrogenation

Product, purity:	i-Butane, 98 wt.%			i-Butene, 39 wt.%		
Raw material, specification:	n-Butane			i-Butane, 98 wt.%		
Process:	BP C_4-isomerisation			UOP Oleflex		
Capacity:	351,000 t/a i-Butane			300,000 t/a i-Butene (39 wt.%)		
Capital investment	10^6 $	10^6 $	$/t · a	10^6 $	10^6 $	$/t · a
Basis	(IV 79)	(III 87)		(III 82)	(III 87)	
Fixed capital	19.9	26.8	76	40.0	40.9	136
Working capital		4.0	11		6.1	20
Total capital investment		30.8	87		47.0	156
Production costs	Units costs	Units/t	Costs $/t		Units/t	Costs $/t
Raw material costs						
n-Butane 45.7 GJ/t	4 $/GJ (1 $/GJ)	1 t	183 (46)			
i-Butane (98 wt.%)	224 $/t (87 $/t)				1.17 t	262 (102)
	(see first stage)					
Catalyst and auxiliaries costs			20			8
Energy costs						
Electricity	7 ¢/kWh	10 kWh	1			
Steam MP 13 bar	14 $/t	0.2 t	3		0.4 t	5
Cooling water	7 ¢/m³	1.5 m³	0		2.7 m³	0
Boiler feed water	1.50 $/m³				2.5 m³	4
Heating gas	4 $/GJ (1 $/GJ)	0.1 GJ	0 (0)		10.5 GJ	42 (10)
Material and energy costs			207 (70)			321 (129)
Labour and overhead costs						
Operating labour	30000 $/my	1/shift	1		3/shift	2
Operating supervision	45000 $/my	0			1	0
Plant overhead	100% excess		1			2
Capital related costs						
Fixed capital related costs	18%		14			25
Interest on working capital	9%		1			2
By-product credit						
H_2 (90%) 119.7 GJ/t	4 $/GJ (1 $/GJ)				−0.02 t	−10 (−2)
C_1-C_3-Hydrocar. 48.2 GJ/t	4 $/GJ (1 $/GJ)				−0.13 t	−25 (−6)
Electricity	7 ¢/kWh				−23 kWh	−2
Steam HP 41 bar	16 $/t				−2.4 t	−38
Total cost of production $/t			224 (87)			277 (112)

Source: [6.29; 6.30; 6.31]

petroleum-producing countries. The world market evidently does not absorb as much of this methanol as some countries, such as Saudi Arabia in particular, expected when the plants were planned [6.27]. The import and customs policy of the countries in which methanol is processed and at the same time, produced, has additionally aggravatcd the marketing problems of the Arabian countries.

The Arabian methanol could probably be more profitably utilized by processing to mixtures with MTBE and/or TBA. The Saudi Arabian Petrochemical is clearly aiming in this direction. Its joint venture with Italian (Enichem) and Finnish (Neste Oy) participation intends to begin production with a 500,000 t/a MTBE plant in 1988 in Al Jubail [6.28].

Table 6.13. Cost of production for MTBE and TBA via dehydrogenated butane

Product, purity:	MTBE, 99 wt.%			TBA, 91 wt.%		
Raw material, specification:	i-Butene, 39 wt.%			i-Butene, 39 wt.%		
Process:	Hüls MTBE			Hüls TBA		
Capacity:	472,500 t/a MTBE			348,890 t/a TBA		
Capital investment	10^6 $	10^6 $	$/t · a	10^6 $	10^6 $	$/t · a
Basis	(III 82)	(III 87)		(III 82)	(III 87)	
Fixed capital	16.5	16.7	35	11.3	11.4	33
Working capital		2.5	5		1.7	5
Total capital investment		19.2	40		13.1	38
Production costs	Unit costs	Units/t	Costs $/t		Units/t	Costs $/t
Raw material costs						
i-Butene (39 wt.%)	277 $/t (112 $/t) (see Table 6.12)	1.62 t	449 (181)		2.51 t	695 (281)
Methanol	130 $/t	0.37 t	48			
Process water	1.50 $/m³				0.78 t	1
Catalyst and auxiliaries costs			1			1
Energy costs						
Electricity	7 ¢/kWh	13 kWh	1		19 kWh	1
Steam MP 10 bar	14 $/t	0.15 t	2		0.56 t	8
LP 3.5 bar	12 $/t	0.75 t	9			
Cooling water	7 ¢/m³	9 m³	1		24 m³	2
Material and energy costs			511 (243)			708 (294)
Labour and overhead costs						
Operating labour	30000 $/my	3/shift	1		3/shift	1
Operating supervision	45000 $/my	1	0		1	0
Plant overhead	100% excess		1			1
Capital related costs						
Fixed capital related costs	18%		6			6
Interest on working capital	9%		1			1
By-product credit						
LPG 45.7 GJ/t	4 $/GJ (1 $/GJ)	−0.99 t	−181		−1.51 t	−276
			(−45)			(−69)
Total cost of production $/t			339 (207)			441 (234)

Source: [6.25]

Tables 6.12 and 6.13 contain cost estimations of the production costs for MTBE and TBA with varying evaluations of the raw materials used, depending on the origin of the butane. The base chemical, consisting essentially of n-butane and coming from a natural gas or refinery gas separating plant, appears in the calculation either with a transfer price corresponding to its local heating value (4 $/GJ), or, alternatively, at its original cost, estimated for obtaining butane in the oil-producing areas of Saudi Arabia (1 $/GJ).

Production costs for i-butane of 224 $/t or 87 $/t are calculated in the first stage (BP C_4-isomerisation in presence of H_2 and organic chlorides) for these alternatives.

According to this calculation the i-butane is to be dehydrogenated in the second stage in the UOP Oleflex process on AlCr-oxide. After one pass the product stream contains 39% i-butene by wt. Hydrogen and heating gas are useful by-products of this stage. The production costs of the i-butene yielded in this second stage amount to 277 or 112 $/t. Subsequently it is converted to MTBE or TBA, according to choice.

Either 472,500 t/a MTBE (99% by wt.) or 348,890 t/a TBA (91% by wt.) can be produced from 300,000 t/a dehydrogenated fraction containing 39% i-butene by wt. In the most favourable case (Saudi Arabia) the production costs are 207 $/t MTBE or 234 $/t TBA. They are thus appreciably higher for both products when the dehydrogenated fraction rather than C_4-cuts or i-butane/propene are used (see Tables 6.10 and 6.11).

The fixed capital requirements in the various cases refer to site conditions in Western Europe. About 50 to 100% higher plant costs must be reckoned with in Arabian countries, so that corresponding amounts must be added to the production costs.

6.1.5.6 Production Costs for SBA

Table 6.14 contains the cost estimation of production costs for SBA made by the Idemitsu process of direct hydration of n-butene, derived from raffinate II. SBA, with production costs of 456 $/t, is evidently too expensive to compete with other components for use in motor fuels.

6.1.6 Evaluation of Profit for High-Octane Components

When car drivers fill up with gasoline they are interested only in the type of fuel which is suitable for their engine, premium or regular. Their decision on purchase depends then on the price. In recent times various mineral oil companies have been trying to exert influence in favour of their products by means of advertising campaigns about the quality of certain fuel additives (e. g. M 2000 of Shell, Formula CE of BP etc.). This has, however, nothing to do with the octane quality. It is immaterial to the drivers which types of component improving octane number are in the gasoline in order to attain the standardised anti-knock quality. They are not even prepared to pay more for a gasoline of octane number higher than the standard because this would be an unnecessary expense. The compression ratio and hence efficiency of the engines are adjusted to the octane number of the standardised types of fuel, even though values are scattered sometimes [6.33].

The decision about the marketing chances of the individual components for octane number improvement thus depends exclusively on the price advantage offered to the

Table 6.14. Cost of production for *sec*-butanol by direct hydration of *n*-butenes in raffinate II with the Idemitsu process

Product, purity:	*sec*-Butanol, 95 wt.%		
Raw material, specification:	Raffinate II, 92 wt.% *n*-Butenes		
Process:	Idemitsu SBA		
Capacity:	40,000 t/a SBA		

Capital investment	10^6 Y (Yen)	10^6 $	$/t · a
Basis	(III 87)	(III 87)	(III 87)
Fixed capital	1800	13.5	337
Working capital		2.0	51
Total capital inv.		15.5	388

Production costs	Unit costs	Units/t	Costs $/t
Raw material costs			
Raffinate II	310 $/t	1.09 t	338
Catalyst and auxiliaries costs			3
Energy costs			
Electricity	7 ¢/kWh	400 kWh	28
Steam MP 15 bar	14 $/t	1.0 t	14
LP 5 bar	12 $/t	4.0 t	48
Cooling water	7 ¢/m^3	300 m^3	21
Boiler feed water	1.50 $/m^3	0.27 m^3	1
Material and energy costs			453
Labour and overhead costs			
Operating labour	30000 $/my	0.5/shift	2
Operating supervision	45000 $/my		
Plant overhead	100% excess		2
Capital related costs			
Fixed capital related costs	18%		61
Interest on working capital	9%		5
By-product credit			
Butanes 45.7 GJ/t	4 $/GJ	−0.26 t	−48
lighter & heavier ends 48.2 GJ/t	4 $/GJ	−0.10 t	−19
Total cost of production $/t			456

Source: [6.32]
Exchange rate: 100 Y ≙ 0.75 $

gasoline producers. The cost differences are due not merely to differences in price or manufacturing costs, which would be in favour of the oxygenates when crude oil prices are high. The cost differences are also influenced by the product efficiencies which have, in addition, to be taken into consideration. Comparisons of profitability are thus needed (see Sect. 6.1.6.2).

6.1.6.1 Comparison of Properties

Both alkylate and polymer gasolines and the components of pyrolysis gasoline possess distinctly lower RON- and MON-efficiencies than the oxygenates. Table 6.15 contains average values for the individual components, obtained separately in test motors.

Table 6.15. Property spectrum of fuel components of high octane number [6.34]

Property	Oxygenates								Hydrocarbons						
	Me-thanol	Ethanol	IPA	SBA	TBA	50/50 TBA/Me-thanol	MTBE	TAME	C$_5$/C$_6$-Iso-merate	C$_3$-Di-meri-sate	Polymer gasoline	Light FCC gasoline	C$_3$-Alky-late	C$_4$-Alky-late	Refor-mate 99
RON	123	121	117	108	106	115	116	108	89	96	100	92	91	94.5	99
MON	91	97	95	91	89	95	98	96	87	82	85	81	88	82.5	88
Upper calorific value [kJ/kg]	22700	27000	33300	35700	35600	29200	38200	39400	ca. 46000 →						
Specific gravity 20 °C [kg/l]	0.793	0.790	0.780	0.800	0.789	0.770	0.741	0.770	0.625	0.690	0.754	0.685	0.690	0.701	0.800
Solubility in water 20 °C [g/100 g Solution]	∞	∞	∞	36.9	∞	∞	1.3	0.6	—	—	—	—	—	—	—
V. P. after Reid [bar]	5.0	3.0	1.5	n. a.	0.48	2*	0.53	n. a.	0.90	0.42	0.14	0.43	0.24	0.24	0.37
Melting point [°C]	−98	−115	−90	−100	+15	−75	−108	n. a.	<−10	<−100	<−100	<−100	<−100	<−100	<−100

* a mixture of 50/50 TBA/MeOH, blended with 4 to 10% by vol. with a gasoline of vapour pressure of 0.70 to 0.90 bar, increases its V. P. by 0.065 to 0.075 bar [6.34]

The refineries do not introduce the components separately but in combinations, thereby optimizing their spectrum of properties. This also yields synergistic effects between components of higher and lower octane number.

The blending octane value obtainable from the components has to be empirically and individually found out for each refinery gasoline pool according to the relation:

$$\text{blending octane value} = \frac{ON - ON_{basis}\ (1-x)}{x}$$

where ON = RON or MON with the added blended component, ON_{basis} = RON or MON of the thereby treated gasoline mixture, x = proportion by volume of the added blending component (by weight in a first approximation).

A disadvantage of the oxygenates (MTBE, TBA and methanol) is that they can be blended with the gasoline only in certain amounts. Field experiments have shown that blends with more than about 20% by vol. of MTBE or 10% by vol. of TBA/methanol, i. e. with oxygen contents of over 3.5% by wt. (Fig. 6.8) are too lean [6.34; 6.35]. Fuels with still higher contents of oxygenates can be used only by vehicles which have a λ-probe for controlling the amount of air for the combustion.

Among the oxygenates, MTBE has a clearly superior anti-knock effect to TBA. This high octane level can be reached only with mixtures of TBA and methanol. MTBE has additional advantages. Their significance in technical application is shown in Table 6.16.

6.1.6.2 Comparison of Profitability

So far it has been shown that the oxygenates MTBE and TBA/methanol are more effective than polymer and alkylate gasoline in improving the octane number. In addition, MTBE can always be produced more economically than polymer gasoline.

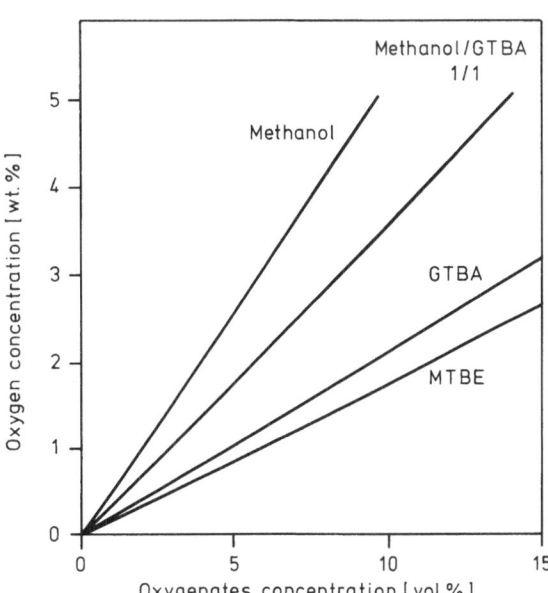

Fig. 6.8. Oxygen content of gasoline plotted against content of various oxygenates [6.36]

Table 6.16. Advantageous properties of MTBE compared to TBA/methanol and their effects in various stages of application

Advantageous property of MTBE	Effect on the application stage		
	refinery	storage tanks/ filling stations	motor vehicle function
higher MON			better engine behaviour at higher speed
higher calorific value			lower gasoline consumption
lower water uptake		water-tight tanks not needed, fewer corrosion problems	no formation of water vapour bubbles during combustion
lower vapour pressure	blending with more butanes is possible		
lower oxygen content	larger amounts can be blended without exceeding legal O$_2$-content limits		
non-corrosive	no addition of inhibitors needed	less corrosion damage*	less wear of tank, fuel pipes, cylinders and exhaust system*
more tolerant of elastomers		tubing on gasoline pumps not attacked	fuel-carrying tubing not attacked

* does not apply to TBA/methanol if rust inhibitors are added

This applies equally to alkylate gasoline, provided that *i*-butene-containing crack-fractions are available in sufficient amounts and are thus as cheap as cuts containing *n*-butenes. TBA/methanol can be produced using the Arco process more cheaply than MTBE and alkylate gasoline, provided that there are enough marketing outlets for the co-product propene oxide.

In the profitability comparison of anti-knock active components, the oxygenates — restricted to the mentioned conditions of their competitiveness – are superior to all the other components which the refineries produce only by themselves (polymer and crack gasoline, dimerisate, alkylate, isomerate and reformate).

The profitability comparison of the components is done here by quantifying the economic profit which the refineries gain when they partly replace their own produced components by oxygenates. The reference value here is the octane number gap which is created by switching production to unleaded gasoline and discontinuing the use of lead-containing additives. The study is made of two model versions assuming the complete substitution of the lead additives:

a) The increased demand for high-octane components is met alone by altering refinery production (cracking, alkylation, isomerisation and, above all, reforming at higher severity).

b) The increased demand is met by partial replacement of the components under a) by oxygenates.

The comparison is simplified when the refineries have large capacity reserves also for producing components of high octane number and the increase in octane number can be accomplished essentially by augmenting the reformer severity without significant extra investment.

The profitability of the lead-free components can be measured in case a) by the extra·consumption of raw materials and energy for making the substitute products. This can be expressed simply as the extra consumption of crude oil. In the reverse sense, saving of crude oil is the measure of the profitability of the oxygenates. This saving of crude oil is achieved by the refineries not having to produce their own components now replaced by oxygenates.

The A. D. Little Institute carried out this substitution calculation for a standard fuel with the optimum economic use of materials and RON value of 94 (see Sect. 6.1.2). Further assumptions of the model and the results are presented in Table 6.17.

For the total substitution of the lead additives in the 17.5 million t gasoline, without any added oxygenates, 578,000 + 35,000 = 613,000 t more crude oil would have to be put into the German refineries in order to obtain a standard gasoline quality of RON = 94 (Table 6.17). If the energy saving is taken into consideration, which arises from not having to make the lead additives, 594,000 t less crude oil would be needed. If the 17.5 million t gasoline of the same quality contain 5% by wt. MTBE (absolute amount of 980,000 t), a saving of 594,000 t crude oil could be achieved and, in addition to this, another 542,000 + 2000 = 544,000 t by lowering the reformer severity. This is a total saving of 1,138,000 t crude oil. Alternatively, the use of 5% by wt. of TBA would save 1,188,000 t crude oil; and the use of 2.5% by wt. of TBA and 1% by wt. methanol together (absolute amount of 623,000 t), a saving of 686,000 t crude oil.

Table 6.17. Estimate of the changes in consumption of raw materials and energy for preparing unleaded gasoline with and without oxygenates (FRG) [6.37]

Basic assumptions: refinery structure of 1981
gasoline production of 17.5 million t
standard quality with RON = 94

Changes in primary energy consumption (in 1000 t crude oil "Arabian Light")	Without oxygenates	5% by wt. MTBE	5% by wt. TBA	3.5% by wt. TBA/methanol*
Crude oil	578	−542	−598	−113
Electric energy	35	−2	4	21
Lead additives	−19	−19	−19	−19
MTBE		816		
TBA			1012	
TBA/CH$_3$OH				393
	594	253	399	282

Conversions: 248 g crude oil/kWh; 5.4 t crude oil/t lead additives; 0.916 t crude oil/t MTBE; 1.136 t crude oil/t TBA; 0.773 t crude oil/t methanol

* weight ratio TBA/methanol = 70/30 (e. g. Oxinol 30 of Arco)

The evaluation of these mentioned savings, based on the current crude oil prices, gives the equivalent values for the corresponding oxygenates. In Fig. 6.9 the profitability of the oxygenates is expressed as a function of crude oil price and dollar rate of

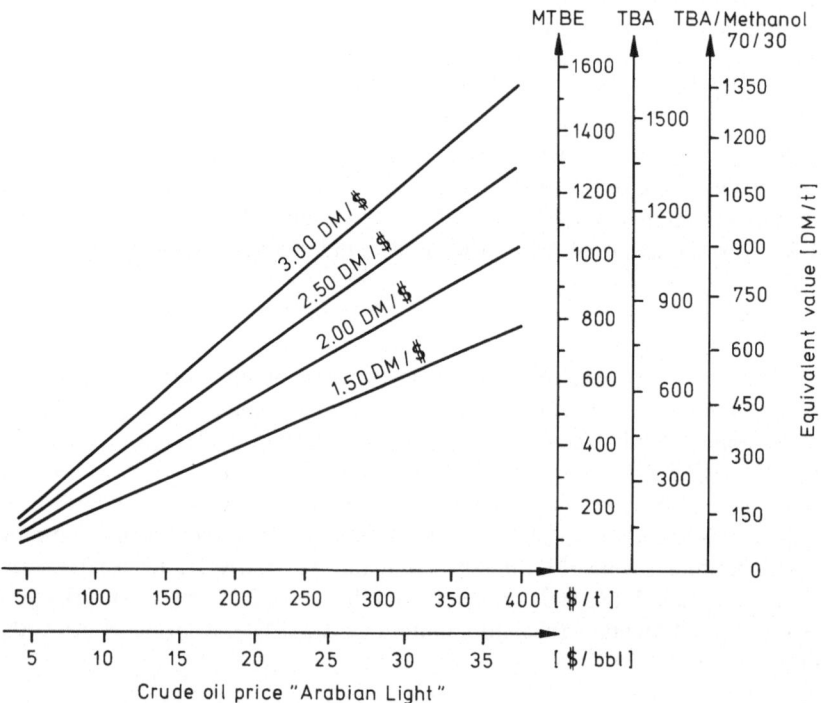

Fig. 6.9. Equivalent value of oxygenates in the FRG in relation to the crude oil price at different DM/$ exchange rates

Table 6.18. Cost of improvement of octane number

Product	RON	MON	ON	Production costs $/t	Production costs/ octane number		
					$/t : RON	$/t : MON	$/t : ON
C$_4$-Alkylate gasoline	94.5	92.5	93.5	300	3.17	3.24	3.21
Polymer gasoline	100	85	92.5	280	2.80	3.29	3.03
MTBE	116	98	107	218	1.88	2.22	2.04
TBA	106	89	97.5	220	2.08	2.47	2.26
GTBA	106	89	97.5	290	2.74	3.26	2.97
SBA	108	91	99.5	456	4.22	5.01	4.58
Methanol	123	91	107	130*	1.06	1.43	1.21
GTBA/methanol (50/50)	115	95	105	210	1.83	2.21	2.00
IPA	117	95	106	400*	3.42	4.21	3.77
Toluene	115	103.5	109.5	200*	1.74	1.93	1.83

* market prices instead of production costs

exchange. There is a basic recognisable tendency towards increased attraction to the use of the oxygenates as crude oil prices rise, and vice versa.

A simplified profitability comparison of the different gasoline components in their ability to increase octane number can also be carried out by relating the production costs to the octane numbers of the components. These last can be expressed as RON, MON or the arithmetic mean, ON = (RON + MON)/2. Table 6.18 contains these data, based on the production costs calculated for all the different components except three (methanol, IPA and toluene) for which the market prices are used. Methanol comes off best here with the lowest values but it must be noted that its unfavourable properties set a limit to its use as a gasoline component. As reference compound for the aromatics, toluene is favourably priced but the refinery structures allow only a limited increase in production. In mixtures with methanol GTBA is as good as MTBE in its ratio of production costs to octane number.

6.1.7 Market Development for Oxygenates

It has been shown above that the oxygenates are cheaper to prepare and also more efficient in their action than other components available for replacing lead alkyls. These favourable conditions for competition gives them the opportunity of filling the gap left in the market by renunciation of the lead alkyls and benzene, and they can also achieve market success at the expense of the other components when crude oil prices are high.

The dominant position of the oxygenate producers, due to the competitive power of their specific products, is reinforced by their monopolistic situation as suppliers of TBA and oligopolistic situation with MTBE. Arco Chemicals is the only world producer of TBA as a motor fuel component; MTBE is made by a limited number of licensed producers and offered freely or produced by some refineries for their own use.

These superior conditions for competition have enabled the oxygenate producers to achieve maximum profits by means of a rigid price-quantity policy, especially in times of higher crude oil prices. Their extremely advantageous market position can presumably continue for only a limited time, for the particular reason that refineries themselves are taking over their own production of MTBE. The market prices of MTBE, in both USA and W. Europe, are tending to increase relative to spot gasoline prices (Fig. 6.10).

Since 1980 legal regulations have existed in the USA for the use of oxygenates in the motor fuel sector. Temporary regulations were in existence in Western Europe for a long time, becoming legally effective at the beginning of 1988 (Table 6.19).

Figure 6.11 shows the real market behaviour of the gasoline suppliers in the use of oxygen-containing components. It can be seen that the independent filling stations offer a gasoline with sometimes distinctly higher content of oxygenates than the filling stations for branded gasoline. In Western Europe in 1985 1.7 million t oxygenates were produced for the gasoline market; of this, 35% were methanol, 35% MTBE, 27% GTBA. The remaining 3% was divided among ethanol, i-propanol, i-butanol, ether and alcohol mixtures, e. g. MAS (methanolo plus alcooli superiori) from the 15,000 t/a plant of Enichem in Pisticci [6.53].

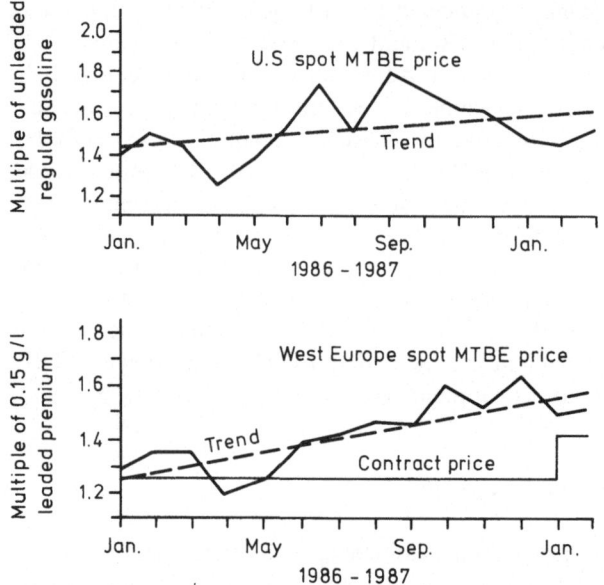

Fig. 6.10. MTBE prices relative to spot gasoline prices in the USA and Western Europe [6.38]

Based on the future legally permitted blended amounts of maximum 10% by vol. MTBE or 7% by vol. TBA*, the present gasoline demand of 24 MMt/a corresponding to 32×10^9 l gives potential markets for a maximum of 2.4 MMt (3.2×10^9 l) MTBE

Table 6.19. Maximum permitted contents of oxygen-containing components of gasoline in W. Europe and the USA

	W. Europe[1]	USA[2]
Max. oxygen content of the gasoline without exceeding the maximum content of the components below:	2.5% by wt.	2% by wt.
Methanol without co-solvent	0% by vol.	0.3% by vol.
Methanol with obligatory co-solvent	3% by vol.	2.75% by vol.
Ethanol	5% by vol.	5% by vol.
i-Propanol (IPA)	5% by vol.	9% by vol.
$tert$-Butanol (TBA)	7% by vol.	11% by vol.
i-Butanol (IBA)	7% by vol.	11% by vol.
Ethers (MTBE, ETBE, TAME)	10% by vol.	11% by vol.
Other monoalcohols	7% by vol.	max. 2% O_2 by wt.

[1] Directive negotiated by the European Fuel Oxygenates Association (EFOA) in cooperation with the Mineral Oil and Auto Industry; approved by the EEC-Commission since Dec. 5, 1985 and come into force as a European Community Regulation on Jan. 1, 1988

[2] US-Regulation of the Environmental Protection Agency (EPA), Federal Register, Vol. 45, pp 67,443 to 67,448 of Oct. 10, 1980 for leaded gasoline, Federal Register, Vol. 46, pp 38,582 to 38,586 of July 28, 1981 for unleaded gasoline

---

* with 3% methanol by vol. the oxygen content of the gasoline is approx. 2.5% by wt.

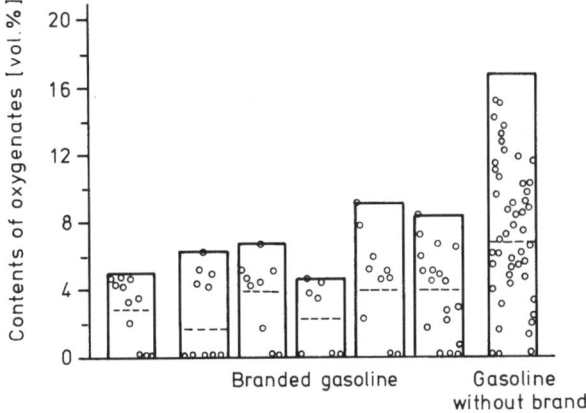

Premium gasoline, leaded and unleaded

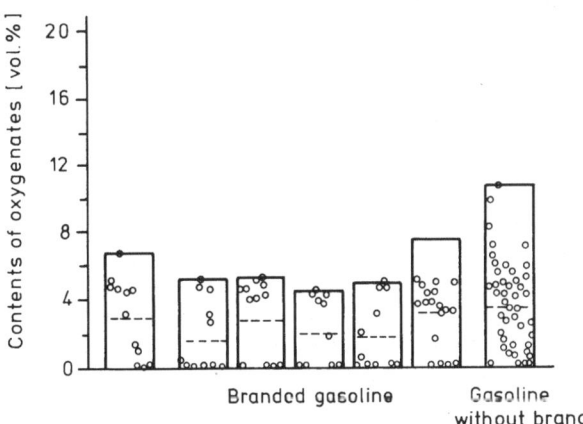

Regular gasoline, leaded and unleaded

Fig. 6.11. Oxygenate contents of various Otto engine fuels in the FRG, 1985 [6.40]

or 1.7 MMt (2.2×10^9 l) TBA. These potential amounts exceed considerably the production capacities for oxygenates at present and in the foreseeable future.

Despite the particularly favourable costs of production of the TBA/methanol mixture and its large contribution to the octane number, Arco is able to exploit the advantages of its monopoly position with limited success. The chances for producing TBA are restricted by the market outlets for the co-product, propene oxide. On the other hand Arco cannot take price-policy advantage of the limitation of supplied amounts of TBA which thereby result, because the MTBE producers are market and price leaders. They have gained market leadership, because MTBE is ahead of TBA in its properties and also there are no supply restrictions associated with a co-product.

TBA has a positive competitive advantage that it can be produced without problems with the supply of raw materials. Methanol and i-butane (if necessary produced by isomerisation of n-butane) will be available in future in sufficient amounts, and, further, the cost advantages of the Arco process permit, in the event of a shortage of propene, its procurement to be financed more readily than by any other propene users. On the other hand the *i*-butene needed to produce MTBE is of only limited availability. If all the refinery B-B and steam cracker C_4-cut in the FRG were used only for MTBE, the maximum amount of MTBE which could be produced is $225,000/0.71 + 337,000/0.64 = 843,000$ t/a (data from Tables 2.6, 2.8 and 4.7).

If the way out of this shortage is sought by making *i*-butene from *i*-butane, the costs of producing the MTBE increase so much that it is more expensive than most of the components produced by the refinery itself. The better alternative is first to convert surplus TBA from the Arco process into MTBE; in this connection Arco has begun a worldwide programme of restructuring their plants which they aim to have completed by the end of 1988. Here again, however, one is tied by the marketing demand for the co-product, propene oxide.

Employment of branched olefines (isoamylenes) of the C_5-fraction have been recently regarded as a way of broadening the raw material basis. *tert*-Amyl methyl ether (TAME) can be made from this with methanol analogously to MTBE derivation, and is of similar octane quality and costs of production. At the same time, this improved utilization of the steam cracker C_5-fraction could contribute to a more economic production of isoprene (see Sect. 3.3.4.4). EC, Cologne-Worringen, developed the TAME process originally for application in the pharmaceutical industry and then adapted it to refinery purposes [6.41]. The first TAME plant (50,000 t/a) from FCC C_5-streams, started production in 1987 at Lindsey Oil refinery, Immingham, with a license from EC [6.42]. Gulf Canada also is granting a license for a similar TAME process [6.43]. The MTBE plant of Texaco in Port Neches, TX is also now able to produce TAME after technical adaptation. If no methanol is available for making MTBE, ethanol may be used as a substitute, yielding ethyl *tert*-butyl ether (ETBE). Further, higher olefinic hydrocarbons, such as are yielded in cracking gasoline, can be etherified to oxygenates. The new Etherol process of BP was developed for this purpose. Its large-scale technical plant of 360,000 t/a capacity, began production in the FRG refinery Vohburg early in 1986 [6.44].

The competition prospects for alkylate gasoline will increase only when this competitive scope of the oxygenates is exhausted. The construction of new production plants for alkylate gasoline would, however, be possible only in local combination with the production of large amounts of refinery B-B from catalytic and perhaps even thermal crackers.

6.2 Prospects for C_4-Technologies in the Plastics Field

6.2.1 C_4-Products in the Plastics Field

The technologies of conversion of C_4-hydrocarbons in the chemical industry have economically surpassing importance which is especially realised in the plastics field. Among these products are primarily thermoplastics but also auxiliaries such as plasticizers, polymerisation initiators and stabilizers (Fig. 6.12).

Evaluation of the chances for C_4-technologies is carried out stepwise from the base chemicals through to the end products (bottom up approach). If in a production stage there are competitors from a process already being operated on a large scale and

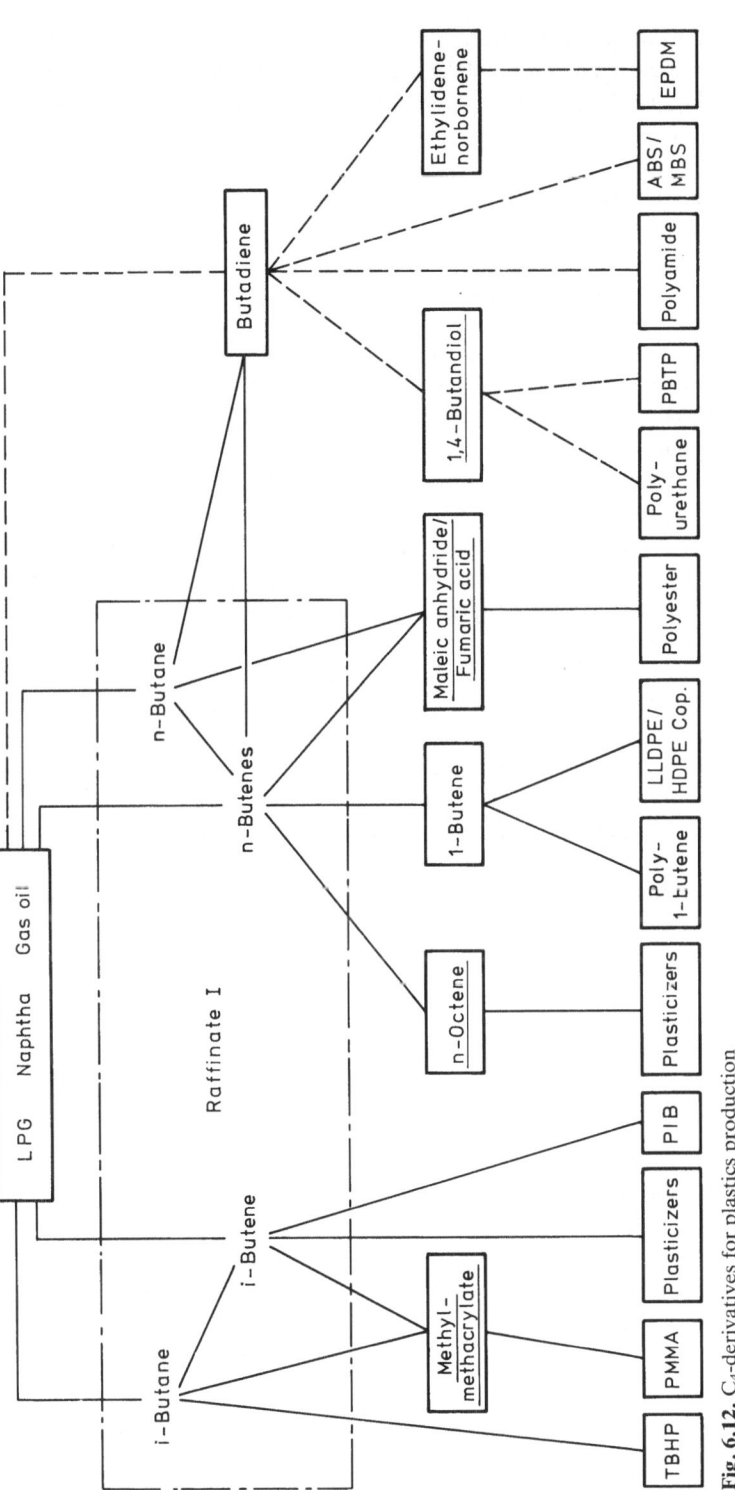

Fig. 6.12. C$_4$-derivatives for plastics production

based on raw materials which do not belong to the C$_4$-family and which furnish products or intermediates of qualitatively equal value (such products or intermediates are underlined in Fig. 6.12), only the production costs will determine the competitive positions of the competing routes. This applies equally to products which are being made by older C$_4$-technologies and for which improved processes are available. If improvements in quality appear thereby this must be assessed separately through a cost-benefit analysis.

Metaeconomic criteria, such as securing the supply of raw materials, handling of dangerous base and intermediate chemicals or ecologically inimical co-products or waste products, can also influence the choice of process.

These potential substitutions in the plastics sector have arisen as a result of engineering advantages in making products on a large scale. Additionally these new C$_4$-chemical products can stimulate existing markets to expand or open new markets based on C$_4$-raw materials.

Especially pure C$_4$-compounds are needed as starting materials for preparing numerous C$_4$-plastic products. Production of these pure compounds is expensive and this can affect decisively the competitive strength of the end products. Consequently the costs of obtaining pure i-butene, 1-butene and butadiene are treated as additional cost burdens. The processing costs are generally calculated as the difference between costs and raw material costs.

There follows a study of the competition for plastics, classified according to their parent butane- and butene-secondary products. There are many kinds of such products and these find a still larger number of end uses. This means that the analysis of competitive conditions, within the scope of the present work, has had to be restricted to some individual examples of particular interest for the comparisons of application benefits. Butane- and butene-secondary products are more rewarding sources for this study than is butadiene. This is because butanes and butenes can have highly contrasting uses, being processed as energy suppliers to motor fuel components and as materials to plastics. It thus comes to particularly diverse types of application benefits, and extensive substitution competition. On the other hand, butadiene serves exclusively for making polymers which are all used as materials (plastics, rubbers) or similar materials (varnishes, paints). Most of the butadiene secondary products have been on the market for a long time and are in established market positions, competing little among themselves.

6.2.2 Costs of Obtaining Pure C$_4$-Base Chemicals

6.2.2.1 Production of Pure i-Butene

i-Butene is accessible via several routes. The choice of route is governed decisively by the raw material situation, the product purity required and the production costs. The results of cost estimates for producing i-butene, given in Tables 6.21 and 6.22, are summarised in Table 6.20.

The molecular sieve extraction yields i-butene purities of 99.1% by wt. at the most. All the other processes give 99.9% purities (by wt.). The sulfuric acid extraction was formerly the only way of obtaining pure i-butene. As a result of environmental

Table 6.20. Production costs and product purities for i-butene in Western Europe, 1986

	Raw material	Process description, Section:	Production costs, Table:	$/t	Processing costs of i-Butene ($/t)	Product purity wt. %
BASF sulfuric acid extraction	raffinate I	3.1.2.2	6.21	478	213	99.9
UCC OlefinSiv molecular sieve extraction	raffinate I	3.1.2.3	6.21	343	91	99.1
Hüls MTBE process and decomposition	raffinate I	3.1.2.4	6.22	330	60	99.9
Hüls TBA direct hydration and decomposition	raffinate I	3.1.2.4	6.22	293	46	99.98
Arco TBA as co-product of propene oxide production, with decomposition and final distillation	i-butane, propene	2.3.2	6.22	300	66	99.9

problems (formation of acid tar, [6.45]) and high production costs it has been largely replaced by more modern processes.

Since it is certain that refineries wish to produce the cheapest possible MTBE and hence tolerate impurities (e. g. dimethyl ether) which can be separated only with difficulty, refinery MTBE from refinery B-B as raw material is unsuitable for decomposition. It has therefore not been considered here.

Although their production costs are practically identical, the TBA routes are preferred today to the MTBE route, because the MTBE decomposition needs a MTBE quality which can only be achieved by a complicated separation process which increases the capital and energy requirements. Hüls TBA contains about 9% by wt. of water after the hydration, but it can be directly subjected to decomposition. Arco TBA contains 5% by wt. of impurities which have to be removed form the product stream; about 20 $/t are added to cover these extra costs.

6.2.2.2 Production of Pure 1-Butene

Among the n-butenes only 1-butene in pure form has so far been suitable for producing chemical products. Table 6.23 summarizes the relevant processes with their production costs. Details of the calculations of these production costs can be found in the following Tables 6.24, 6.25 and 6.26.

In all cases the purity exceeds 99% 1-butene by weight. i-Butene is completely removed by the MTBE process and butadiene by selective hydrogenation.

6.2.2.3 Production of Pure Butadiene

The C$_4$-cuts from ethylene steam crackers are the sole source of butadiene in Western Europe and make up over 94% of world production. The extraction processes used have been described in Sect. 3.1.1.

Table 6.21. Cost of production for pure i-butene by acid extraction and by molecular sieve extraction

Product, purity:	i-Butene, 99.9 wt.%			i-Butene, 99.1 wt.%		
Raw material, specification:	Raffinate I, 50 wt.% i-Butene			Raffinate I, 45 wt.% i-Butene		
Process:	BASF H_2SO_4 Extraction			UCC OlefinSiv		
Capacity:	100,000 t/a i-Butene			100,000 t/a i-Butene		
Capital investment	10^6 $	10^6 $	$/t · a	10^6 $	10^6 $	$/t · a
Basis	(II 80)	(III 87)		(III 77)	(III 87)	
Fixed capital	36.7	45.1	451	7.7	12.1	121
Working capital		6.8	68		1.8	18
Total capital investment		51.9	519		13.9	139
Production costs	Unit costs	Units/t	Costs $/t		Units/t	Costs $/t
Raw material costs						
Raffinate I	280 $/t	2.22 t	622			
(50 wt.% i-Butene)						
Raffinate I	280 $/t				2.47 t	692
(45 wt.% i-Butene)						
Catalyst and H_2SO_4 (98%)	10 ¢/kg	4 kg	0			
auxiliaries NaOH (98%)	30 ¢/kg	3 kg	1			
costs Nitrogen	35 $/1000 m³$_n$				50 m³$_n$	2
Energy costs						
Electricity	7 ¢/kWh	500 kWh	35		110 kWh	8
Steam LP 4 bar	12 $/t	2.50 t	30		3.15 t	38
Cooling water	7 ¢/m³	400 m³	28		80 m³	6
Heating gas	4 $/GJ	4.9 GJ	20		0.3 GJ	1
Material and energy costs			736			747
Labour and overhead costs						
Operating labour	30000 $/my	3/shift	5		3/shift	5
Operating supervision	45000 $/my	1	1		1	1
Plant overhead	100% excess		6			6
Capital related costs						
Fixed capital related costs	18%		81			22
Interest on working capital	9%		6			2
By-product credit						
Raffinate II	310 $/t	−1.15 t	−357		−1.42 t	−440
C_8+-Isomers	−	−0.05 t	0		−0.03 t	0
Total cost of production $/t			478			343

Sources: [6.46; 6.47; 6.48; 6.49]

It is hardly possible to make statements of general validity about the economic advantages of any particular process. Factors of importance are the type of process, the degree of modernization of the plant, the infrastructure, the license fees, and the specifications of raw materials and product. The processes are being continuously adapted to the particular energy costs. Prominent among new plants are the process of BASF and Nippon Zeon, for which the comparative production costs on a standard basis are given in Table 6.27.

The cost estimation is based on versions of the process in which a part (not given) of the electricity is obtained from the high pressure steam supplied (e. g. from steam

Table 6.22. Cost of production for pure *i*-butene by MTBE and TBA decomposition

Product, purity:	*i*-Butene, 99.9 wt.%			*i*-Butene, 99.9 wt.%			
Raw material, specification:	MTBE, 99 wt.%			TBA, 99 wt.% (GTBA, 95 wt.%)			
Process:	Hüls MTBE Decomposition			Hüls (Arco) TBA Decomposition			
Capacity:	100,000 t/a *i*-Butene			100,000 t/a *i*-Butene			
Capital investment	10^6 $	10^6 $	$/t · a	10^6 $	10^6 $	$/t · a	
Basis	(III 82)	(III 87)		(III 82)	(III 87)		
Fixed capital	5.5	5.6	56	3.3	3.4	34	
Working capital		0.8	8		0.5	5	
Total capital investment		6.4	64		3.9	39	
Production costs	Unit costs	Units/t	Costs $/t		Units/t	Costs $/t	
Raw material costs							
MTBE	218 $/t (see Table 6.10)	1.58 t	344				
TBA	170 $/t				1.47 t	250	
(GTBA)	(182 $/t) (see Table 6.11)				(1.41 t)	(257)	
Catalyst and auxiliaries costs			1				
Energy costs							
Electricity	7 ¢/kWh	20 kWh	1				
Steam MP 15 bar	14 $/t	0.3 t	4				
LP 5 bar	12 $/t	2.2 t	26		1.9 t	23	
Cooling water	7 ¢/m^3	75 m^3	5		55 m^3	4	
Material and energy costs			381			277 (284)	
Labour and overhead costs							
Operating labour	30000 $/my	3/shift	5		3/shift	5	
Operating supervision	45000 $/my	1	1		1	1	
Plant overhead	100% excess		6			6	
Capital related costs							
Fixed capital related costs	18%		10			6	
Interest on working capital	9%		1			1	
By-product credit							
Methanol	130 $/t	−0.57 t	−74				
TBA	170 $/t (182 $/t)				−0.01 t	−2	
Dimethyl ether	−	−5 kg	0		−	−	
C$_8$+-Isomers	−	−5 kg	0		−2 kg	0	
Process water	1.50 $/m^3				−0.45 m^3	1	
Total cost of production $/t			330			293 (300)	

Source: [6.25]

crackers in the vicinity). The extraction agent NMP contains 5 to 10% water by wt., which improves the selectivity of butadiene extraction and lowers its boiling point. DMF is highly selective for butadiene. Addition of water to lower the boiling point would be welcome but this is not done because it would induce corrosion. Therefore a higher temperature in the stripper is applied, causing a larger consumption of steam and necessitating addition of a polymerisation inhibitor such as 4-*tert*-butylpyro-

Table 6.23. Production costs and product purities for 1-butene in Western Europe, 1986

Processes	Raw material	Process description, Section:	Production costs, Table:	$/t	Processing costs of 1-Butene ($/t)	Product purity wt.%
Distillation*	raffinate II	3.1.2.2	6.24	496	42	99.7
IFP Alphabutol ethylene dimerisation	ethylene	2.2.4.1	6.24	496	65	99.8
UOP Sorbutene molecular sieve extraction	raffinate I	3.1.2.4	6.25	361	81	99.2
UOP Sorbutene molecular sieve extraction	catcracker C_4-cut	3.1.2.4	6.25	453	270	99.2
n-Butane dehydrogenation with UOP Sorbutene molecular sieve extraction	n-butane	3.2.2, 3.1.2.4	6.26	472	146	99.2

* corresponds roughly to obtaining 1-butene from the residual fraction of the α-olefine preparation (Sect. 2.2.4.1)

Table 6.24. Cost of production for 1-butene by raffinate II fractionation and ethylene dimerisation

Product, purity:	1-Butene, 99.7 wt.%			1-Butene, 99.8 wt.%		
Raw material, specification:	Raffinate II, 50 wt.% 1-Butene			Ethylene		
Process:	Fractionation			IFP Alphabutol		
Capacity:	57,000 t/a 1-Butene			50,000 t/a 1-Butene		
Capital investment	10^6 $	10^6 $	$/t · a	10^6 $	10^6 $	$/t · a
Basis	(II 77)	(III 87)		(III 85)	(III 87)	
Fixed capital	5.1	8.0	140	3.8	3.8	76
Working capital		1.2	21		0.6	12
Total capital investment		9.2	161		4.4	88
Production costs	Unit costs	Units/t	Costs $/t		Units/t	Costs $/t
Raw material costs						
Raffinate II	310 $/t	2.13 t	661			
Ethylene	410 $/t				1.05 t	431
Catalyst and auxiliaries costs						11
Energy costs						
Electricity	7 ¢/kWh	3 kWh	0		19 kWh	1
Steam MP 14 bar	14 $/t				0.75 t	11
LP 4 bar	12 $/t	0.67 t	8			
Cooling water	7 ¢/m³	20 m³	1		94 m³	7
Material and energy costs			670			461
Labour and overhead costs						
Operating labour	30000 $/my	1/shift	3		3/shift	9
Operating supervision	45000 $/my	—			1	1
Plant overhead	100% excess		3			10
Capital related costs						
Fixed capital related costs	18%		25			14
Interest on working capital	9%		2			1
By-product credits						
2-Butene, Butanes 45.7 GJ/t	4 $/GJ	−1.13 t	−207			
Hexenes	—				−0.05 t	0
Total cost of production $/t			496			496

Sources: [6.50; 6.51]

Table 6.25. Cost of production for 1-butene through molecular sieve extraction of steam cracker- and catcracker-C_4-cut

Product, purity:	1-Butene, 99.2 wt.%			1-Butene, 99.2 wt.%		
Raw material, specification:	Raffinate I, 26 wt.% 1-Butene			FCC C_4-Cut, 12 wt.% 1-Butene		
Process:	UOP Sorbutene			UOP Sorbutene		
Capacity:	45,000 t/a 1-Butene			45,000 t/a 1-Butene		
Capital investment	10^6 $	10^6 $	$/t · a	10^6 $	10^6 $	$/t · a
Basis	(II 80)	(III 87)		(II 80)	(III 87)	
Fixed capital	22.0	27.0	600	33.0	40.6	902
Working capital		4.0	89		6.1	135
Total capital investment		31.0	689		46.7	1037
Production costs	Unit costs	Units/t	Costs $/t		Units/t	Costs $/t
Raw material costs						
Raffinate I	280 $/t	3.89 t	1089			
FCC C_4-Cut 45.7 GJ/t	4 $/GJ				7.91 t	1446
Catalyst and auxiliaries costs			4			8
Energy costs						
Electricity	7 ¢/kWh	48 kWh	3		80 kWh	6
Steam MP 15 bar	14 $/t	2.20 t	31		3.70 t	52
Cooling water	7 ¢/m^3	69 m^3	5		114 m^3	8
Material and energy costs			1132			1520
Labour and overhead costs						
Operating labour	30000 $/my	3/shift	10		3/shift	10
Operating supervision	45000 $/my	1	1		1	1
Plant overhead	100% excess		11			11
Capital related costs						
Fixed capital related costs	18%		108			162
Interest on working capital	9%		8			12
By-product credits						
Raffinate I,						
1-Butene extracted	280 $/t	−2.89 t	−809			
LPG 45.7 GJ/t	4 $/GJ				−6.91 t	−1263
Total cost of production $/t			361			453

Source: [6.52]

catechol or 2,6-di-*tert*-butyl-*p*-cresol as *i*-butene specialities. The initiator must be washed out of the product with NaOH at the end.

The production costs for butadiene are about 700 $/t for the process with the assumptions made. The BASF process has higher capital costs, the Nippon Zeon process has higher energy costs.

Table 6.26. Cost of production for 1-butene via butane dehydrogenation with molecular sieve extraction

Product, purity:	n-Butenes, 99 wt.%			1-Butene, 99.2 wt.%		
Raw material, specification:	n-Butane			n-Butenes, 93 wt.%		
Process:	UOP Oleflex with Olex			UOP Sorbutene		
Capacity:	290,000 t/a n-Butenes			100,000 t/a 1-Butene		
Capital investment	10^6 $	10^6 $	$/t · a	10^6 $	10^6 $	$/t · a
Basis	(III 82)	(III 87)		(II 80)	(III 87)	
Fixed capital	54.2	55.4	191	45.4	55.8	558
Working capital		8.3	29		8.4	84
Total capital investment		63.7	220		64.2	642
Production costs	Unit costs	Units/t	Costs $/t		Units/t	Costs $/t
Raw material costs						
n-Butane 45.7 GJ/t	4 $/GJ	1.24 t	227 (57)			
	(1 $/GJ)					
n-Butenes	232 $/t (102 $/t)				2.90 t	673 (296)
(35 wt.% 1-Butene)	(see first stage)					
Catalyst and auxiliaries costs			8			2
Energy costs						
Electricity	7 ¢/kWh				25 kWh	2
Steam MP 15 bar	14 $/t				1.50 t	21
LP 3.5 bar	12 $/t	0.40 t	5			
Cooling water	7 ¢/m³	3 m³	0		40 m³	3
Boiler feed water	1.50 $/m³	2.8 m³	4			
Heating gas	4 $/GJ	11,7 GJ	47			
Material and energy costs			291 (121)			701 (324)
Labour and overhead costs						
Operating labour	30000 $/my	3/shift	2		3/shift	5
Operating supervision	45000 $/my	1	0		1	0
Plant overhead	100% excess		2			5
Capital related costs						
Fixed capital related costs	18%		34			100
Interest on working capital	9%		3			8
By-product credit						
Electricity	7 ¢kWh	−27 kWh	−2			
H$_2$ (85%) 119.7 GJ/t	4 $/GJ	−0.03 t	−14 (−4)			
	(1 $/GJ)					
C$_1$-C$_3$-Hydroc. 48.2 GJ/t	4 $/GJ	−0.21 t	−40 (−10)			
	(1 $/GJ)					
Steam HP 41 bar	16 $/t	−2.70 t	−43			
Condensate	1.50 $/m³	−0.4 m³	−1			
LPG 45.7 GJ/t	4 $/GJ				−1.90 t	−347 (−87)
	(1 $/GJ)					
Total cost of production $/t			232 (102)			472 (355)

Sources: [6.25; 6.52]

Table 6.27. Cost of production for 1,3-butadiene by C$_4$-cut extraction

Product, purity:	1,3-Butadiene, 99.8 wt.%			1,3-Butadiene, 99.8 wt.%		
Raw material, specification:	C$_4$-Cut, 45 wt.% Butadiene			C$_4$-Cut, 45 wt.% Butadiene		
Process:	BASF NMP Extraction			Nippon Zeon DMF Extraction		
Capacity:	50,000 t/a Butadiene			50,000 t/a Butadiene		
Capital investment	10^6 $	10^6 $	$/t · a	10^6 $	10^6 $	$/t · a
Basis	(III 77)	(III 87)		(III 77)	(III 87)	
Fixed capital	11.9	18.7	374	10.1	15.9	318
Working capital		2.8	56		2.4	48
Total capital investment		21.5	430		18.3	366
Production costs	Unit costs	Units/t	Costs $/t		Units/t	Costs $/t
Raw material costs						
C$_4$-Cut	400 $/t	2.29 t	916		2.29 t	916
(2.5 wt.% C$_4$-Acetylenics)						
Catalyst and NMP	3 $/kg	0.2 kg	1			
auxiliaries DMF with	4 $/kg				0.2 kg	1
costs inhibitor						
Energy costs						
Electricity	7 ¢/kWh	33 kWh	2		22 kWh	2
Steam HP 42 bar	16 $/t	2.0 t	32		2.7 t	43
Cooling water	7 ¢/m^3	150 m^3	11		175 m^3	12
Material and energy costs			962			974
Labour and overhead costs						
Operating labour	30000 $/my	3/shift	9		3/shift	9
Operating supervision	45000 $/my	1	1		1	1
Plant overhead	100% excess		10			10
Capital related costs						
Fixed capital related costs	18%		67			57
Interest on working capital	9%		5			4
By-product credit						
Raffinate I	280 $/t	−1.26 t	−353		−1.26 t	−353
Total cost of production $/t			701			702

Sources: [6.53; 6.54; 6.55]

6.2.3 Butane Secondary Products in the Plastics Field

6.2.3.1 Maleic Anhydride and Fumaric Acid

Maleic anhydride (MA) is an intermediate which finds wide use. It can be prepared from three different starting materials — benzene, *n*-butane or raffinate II, which can also contain butadiene. Fumaric acid (FA) is, at present, a secondary product of MA and co-product from the preparation of phthalic anhydride (PA) and of MA from non-C$_4$-raw materials.

At present in the USA, where most of the MA and FA are produced, there is a capacity of just 250,000 t/a MA, approx. 70% of which comes from *n*-butane and 30% from benzene. The variation with time (1968−1983) of the production amounts and sales values for MA and FA in this region is shown in Fig. 6.22.

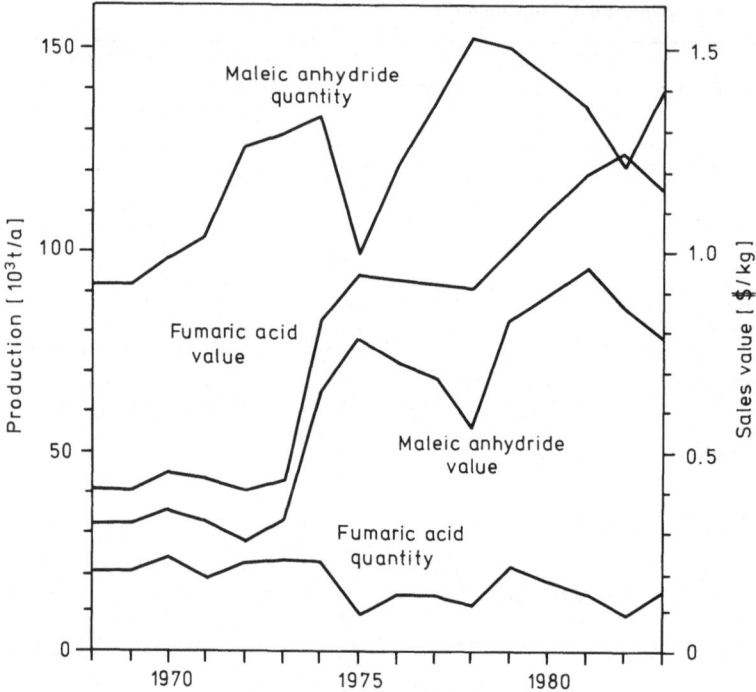

Fig. 6.13. Development of the market for MA and FA in the USA between 1968 and 1983 [6.56]

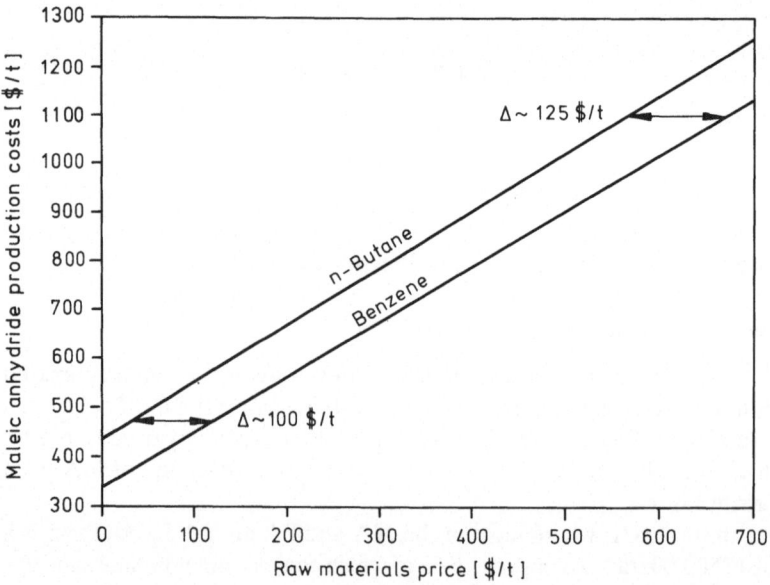

Fig. 6.14. MA production costs for the Halcon/SD benzene and butane processes at various raw material prices

Table 6.28 contains a comparison of the production costs for MA in the Halcon/SD process in its butane- and benzene raw material versions. This is the most used process in Western Europe. The capacity chosen refers to the largest technically feasible size of reactor for the butane process (solid bed reactor with about 25,000 tubes containing the catalyst and surrounded by molten salt for heat removal). Since the turnover is smaller in the butane procedure, the costs for energy and those depending on fixed capital are higher. This could be more than compensated during the last 15 years though lower raw material costs on sites with relatively low butane prices.

The situation is characterised at the moment by uncertainty about the benzene price. The obligation to prepare gasoline free of benzene (see Sect. 6.1.2) could lead

Table 6.28. Cost of production for maleic anhydride from butane or benzene

Product, purity:	MA			MA		
Raw material, specification:	n-Butane			Benzene		
Process:	Halcon/SD-Butane			Halcon/SD-Benzene		
Capacity:	27,200 t/a MA			27,200 t/a MA		
Capital investment	10^6 $	10^6 $	$/t · a	10^6 $	10^6 $	$/t · a
Basis	(IV 80)	(III 87)		(IV 80)	(III 87)	
Fixed capital	17	20.9	768	14	17.2	632
Working capital		3.1	114		2.6	85
Total capital investment		24.0	882		19.8	717
Production costs	Unit costs	Units/t	Costs $/t		Units/t	Costs $/t
Raw material costs						
n-Butane RON = 93	198 $/t	1.17 t	232			
(45.7 GJ/t)	(4 $/GJ)		(214)			
Benzene RON = 108	381 $/t				1.14 t	434
(40.4 GJ/t)	(4 $/GJ)					(184)
Process water	1.50 $/m^3	0.5 m^3	1		0.4 t	1
Catalyst and			60			30
auxiliaries costs Nitrogen	35 $/1000 m$^3_n$	12 m^3_n	0		12 m^3_n	0
Energy costs						
Electricity	7 ¢/kWh	1859 kWh	130		1100 kWh	77
Cooling water	7 ¢/m^3	229 m^3	16		168 m^3	12
Boiler feed water	1.50 $/m^3	12.3 m^3	18		10.6 m^3	16
Material and energy costs			457 (439)			570 (320)
Labour and overhead costs						
Operating labour	30000 $/my	14/shift	77		14/shift	77
Operating supervision	45000 $/my	1	2		1	2
Plant overhead	100% excess		79			79
Capital related costs						
Fixed capital related costs	18%		138			114
Interest on working capital	9%		10			8
By-product credit						
Steam MP 25 bar	14 $/t	−6.6 t	−92		−5.8 t	−81
Condensate	1.50 $/m^3	−4.4 m^3	−7		−3.9 m^3	−6
Total cost of production $/t			664 (646)			763 (513)

Source: [6.57]

to a surplus of benzene on the market with consequent fall in price. Because of this uncertainty, the market prices of the two raw materials are not used in the cost estimation but, instead, their upper and lower prices are used as a motor fuel (see Sect. 6.1.3.2 for the method of evaluation) and when used as an ordinary liquid fuel (calorific value). As is seen in Fig. 6.14, n-butane is competitive as raw material only when its price is 100 to 125 \$/t below that of the benzene.

The argument sometimes advanced [6.58] that benzene emissions during production could lead to a legal ban on new or existing benzene MA plants is not conclusive since, meanwhile, efficient systems for control and prevention of emissions are now features of modern technique.

The data given up to now have been concerned with conventional technologies, realised on the large scale. Fully continuous processing with favourable production costs are now available for the raw material n-butane (see Sect. 3.2.3); however, it has not yet been possible to confirm their economic success on the large scale. The relative cost advantages when assuming 18% larger capacity for FA compared with MA, have recently been published (Table 6.29) [6.56; 6.59].

Table 6.29. Relative production costs for different MA and FA processes

Technology for raw material n-butane	Relative production costs in % (based on MA = 100)		
	MA	FA	FA reduced to the wt. proportion of MA (= FA × 116/98)
Halcon/SD solid bed	100.0	72.5	85.8
Alusuisse Italia solid bed with anhydrous recovery	87.0	63.8	75.5
ALMA fluid bed with anhydrous recovery	75.4	55.1	65.2

The cost advantages for FA are especially notable, due to the lower fixed capital-related costs. The investment costs of producing FA on the benzene basis are higher because it is more complicated to extract the FA on account of formation of impurities such as benzoquinones [6.56].

Up to now FA has been used exclusively, for economic reasons, in only small amounts for making unsaturated polyester resins of especially high quality. Unsaturated polyesters (UP) are the most important derivatives of MA in Western Europe and the USA, with a proportion of about 50% [6.56]. The direct base chemical precursor of UP resins is always a so-called pre-condensate which is processed usually with styrene (sometimes with MMA as comonomer) and with added fibre reinforcement and, if necessary, filler and pigments. Reaction is carried out using a curing agent (at higher temperature in thermo-setting), a curing agent plus accelerator (cold hardening) or photo resistants (light hardening), to yield sheets, moulded parts, vessels or tubes. Another extensive application of unsaturated polyesters is for furniture varnishes.

The pre-condensate is obtained from MA or FA and diols in the melt at 170 to 200 °C, according to the equations:

$$n \begin{matrix} HC-C \\ \| \quad \rangle O \\ HC-C \\ \| \\ O \end{matrix} + n\,HO-(CH_2)_x-OH \xrightarrow[-2n\,H_2O]{-n\,H_2O} \cancel{}$$

MA   Diol $\boxed{}$OOC-CH=CH-COO-(CH$_2$)$_x$$\boxed{}_n$ Pre-condensate for UP-resins

HOOC-CH=CH-COOH + n HO-(CH$_2$)$_x$-OH

FA   Diol

Diols $\begin{cases} x = 2 : \text{Ethylene glycol} \\ x = 3 : \text{Propylene glycol} \\ x = 4 : \text{1,4-Butanediol} \end{cases}$

Using modern *n*-butane processes it would be possible for the first time to prepare FA directly so that production costs are lower than those for MA. UP resins based on FA are generally of superior quality and this could lead to replacement of the MA-UP resins and possible substitution in fields where products have been hitherto made from duroplasts and casein formaldehyde resin. For these applications, the following questions are crucial:

— what is the impact on the UP resins of an FA cost reduction, if completely transferred to the price, and is it enough to be able to undercut continually the critical price threshold of competing products?
— to what extent could this price reduction in combination with the improved quality of FA-UP resins compared to those from MA, lead to replacement of higher grade competing products?

The large number of different resin types makes it difficult to provide a general opinion on how markedly cost reductions of base chemicals affect production costs of the end products. For instance, the Hüls UP resin assortment (all Vestopal types) contains on the average [6.60] 35% styrene, 28% acids (of which 46% PA, 40% MA, 14% FA), 27% glycols, and 10% other components, principally fire-resistant Al(OH)$_3$. In types of product for further processing to resin-impregnated fibre mats (so-called prepregs) there is a tendency to use more MA/FS in the acid part. Types requiring fillers (e. g. marble) for further processing tend to have more PA as acid part as this is a cheaper component.

About 67,000 t/a UP resins and varnishes were produced in 1986 in the FRG. The total capacity of 143,000 t/a was thus only utilized to 46%. The individual capacities and product trade names of the producers are: BASF 80,000 t/a Palatal, Bayer 35,000 t/a Leguval, Hüls 18,000 t/a Vestopal, Hoechst 10,000 t/a Alpolit.

The production in 1985 was distributed among the following market sectors [6.60]: building (slabs, resin concrete, 25%), transport (cars, lorries, rail, airplanes, 23%), electrical industry (12%), vessels and tubes (10%), boats (5%), miscellaneous (plastic fillers, buttons, 25%). Experts believe that the best substitution chances are in the transport and electrical fields.

UP resins are resistant to corrosion and deformation, are light-weight and require little maintenance. This makes them particularly suitable in the auto industry, e. g. as low located spoilers, resistant to mechanical destruction and optimally streamlined.

Studies by Du Pont show that during the lifetime of a motor vehicle, taken as covering 160,000 km or 100,000 miles, 1 kg less weight can save 9 l of gasoline [6.61].

The marketing prospects of the polyester resins are improved by development of rational engineering techniques such as BMC (bulk moulding compound) or SMC (sheet moulding compound) and of new types of fibre (e. g. Aramid fibres), which are lighter than fibre glass although very costly to produce (fibre glass 2.50 $/t, Aramid fibres 60 $/t).

At present the largest amounts of UP resin used in the transport sector are for constructing lorries, caravans etc. In second place is the interior equipment of private cars (sides and roof interior coverings). Complete car bodies (Porsche 959) or mass-produced parts such as boot covers (BMW M3) or bumpers (large Audi limousine) are still exceptions. It has been calculated that in ten years there will be a maximum of 10 kg SMC per car, i. e. if 4 million cars are built in the FRG in a year they will consume an additional amount of about 40,000 t/a SMC or 10,000 t/a UP resins [6.60].

Gradual replacement of the phenol resin moulded material by UP resins (25% resin fraction) is to be expected in the electrical field because of the better dielectric and mechanical properties of the latter. This applies also to electro-constructional parts although not for switch cabinets. These are better made from thermoplasts because of their low processing costs.

In the building industry, UP resins had considerable market success in the 1960s and 1970s but have subsequently been displaced from various fields of application by thermoplasts (corrugated sheets of MMA and PC, oil tanks of PE) or epoxy resins (sealing materials, adhesive mortar). This displacement competition has now been largely concluded, so that in the future in this branch neither further sales losses nor re-substitution can be expected, apart from the influence of ups and downs in the economic situation. The conclusions below may be made:

— the raw material n-butane offers no special advantages for making MA, except under special conditions,
— the butane basis would offer appreciably more favourable costs for making FA, although halving the production costs of FA would reduce the material cost of SMC parts or moulding compounds by only about 3.5%. This value is given by the calculation: 25% resin fraction × 28% FA content. In practice it is still smaller because certain amounts of PA are always mixed with FA. Moreover, the engineering of FA processing is more demanding and hence expensive (FA-"dustcarts" instead of pumped MA), and the product quality is not so markedly improved that entry could be made into, above all, the field of application of the epoxy resins. Consequently the larger polyester processors scarcely see any advantages in going over completely from MA polyesters to the use of FA polyesters.

6.2.3.2 Other Secondary Products

The lower prices and fewer restrictions of availability of the paraffins encourage the chemical industry generally to try to use them as raw materials instead of the corresponding olefines. There have been more developments in C_4-chemistry (see Sect. 3.2) than in C_2- or C_3-chemistry, although most processes have not yet been brought to commercial application. This is true for the production of 1,4-butanediol from *n*-butane via MA and likewise for the derivation of MMA from *i*-butane, for which there are no data available for evaluating production costs and product quality. Apart from this, the absence of an infrastructure for butane chemistry impedes the technological breakthrough. A butane pipeline distribution system, similar to the ethylene pipeline networks existing in Western Europe and in the south of the USA, or even the flexible utilization of this or that pipeline for butane transport, would give the C_4-paraffins a new impetus.

6.2.4 Butene Secondary Products in the Plastics Field

6.2.4.1 Methyl Methacrylate

The production of methyl methacrylate (MMA) by the classical acetone cyanohydrin (ACH) process yields troublesome co-products. Search has thus been proceeding for a long time for processes which are more suitable in this connection (see Sects. 3.2.3 and 3.3.4.3). These start from the raw materials ethylene, propene, acetone, TBA/*i*-butene and *i*-butane, in competition among each other. Large-scale plants have been built only for the ACH process and, quite recently, also for the TBA/*i*-butene route. Only general and subjective estimates of costs can be found in the literature for production based on the other raw materials mentioned. These procedures have not yet reached even the pilot plant stage and there is thus no basis for precise calculations of costs. Among these other starting materials, ethylene has sometimes been in favour [6.62]. According to others, *i*-butane is supposed to be the best raw material [6.63]. Hence the comparison of production costs for MMA made below is restricted to the ACH process of Röhm and the new TBA process of Mitsubishi Rayon.

The TBA route shows a slight cost advantage (Table 6.30) which arises only through the lower demand for capital outlay. The higher capital invested in the ACH process is due to adjoining plants for the production of hydrocyanic acid and for decomposition of the residues to SO_2 for making sulfuric acid. The hydrocyanic acid plants must be kept ready for use in case the external supply of the acid should stop. The labour and capital costs of the hydrocyanic acid plant are fixed costs and must be added to the manufacturing costs of ACH, even when HCN is obtained externally at a market price of 400 $/t. It can be recognised that the economical performance of a depreciated ACH plant is attained only when the TBA plant is also depreciated and the TBA utilized is evaluated at its production cost (see Table 6.11). However, in times of high crude oil prices the use of TBA in the chemical industry will be hampered by its high potential profit in the motor fuel sector (see Fig. 6.9).

The USA is at the forefront of world MMA producers, with a capacity of 540,000 t/a, entirely based on the classical process (see Table 3.10). The amounts produced annually are shown in Fig. 6.15.

This MMA produced in the USA is processed to the products listed in Table 6.31 [6.63].

The traditional markets for MMA in the 1960s and 1970s, with yearly growth rates of 6 to 8% were in the building sector (bath-tubs, shower basins, lighting, corrugated and support sheets, roof covers) and the auto sector (rear lights, surface coatings of acrylate resins). As the growth rate of the building industry decreased and smaller cars

Table 6.30. Cost of production for methyl methacrylate by acetone cyanohydrin process and by TBA oxidation

Product, purity:	MMA			MMA		
Raw material, specification:	Acetone, Hydrogen cyanide, Methanol			TBA, Oxygen, Methanol		
Process:	Röhm ACH			Mitsubishi Rayon TBA Oxidation		
Capacity:	136,000 t/a MMA			136,000 t/a MMA		
Capital investment	10^6 $	10^6 $	$/t · a	10^6 $	10^6 $	$/t · a
Basis	(IV 85)	(III 87)		(IV 85)	(III 87)	
Fixed capital	125.0	123.3	907	90.0	88.8	653
Working capital		18.5	136		13.3	98
Total capital investment		141.8	1043		102.1	751
Production costs	Unit costs	Units/t	Costs $/t		Units/t	Costs $/t
Raw material costs						
Acetone	410 $/t	0.66 t	271			
Hydrogen cyanide	400 $/t	0.29 t	116			
Methanol	130 $/t	0.34 t	44		0.34 t	44
H_2SO_4	*	0.22 t				
TBA (see Table 6.11)	170 $/t (290 $/t)				1.09 t	185 (316)
Oxygen	40 $/t				1.04 t	42
Catalyst and auxiliaries costs						
Energy costs						
Electricity	7 ¢/kWh	530 kWh	37		2895 kWh	203
Steam MP 15 bar	14 $/t	6.4 t	90		6.3 t	88
Cooling water	7 ¢/m³	716 m³	50		841 m³	59
Heating gas	4 $/GJ	9.3 GJ	37		5.8 GJ	23
Material and energy costs			645			644 (657)
Labour and overhead costs						
Operating labour	30000 $/my	10/shift	11		10/shift	11
Operating supervision	45000 $/my	1	0		1	0
Plant overhead	100% excess		11			11
Capital related costs						
Fixed capital related costs	18%		163			118
Interest on working capital	9%		12			9
By-product credit						
Acetic acid	50 ¢/kg				−75 kg	−38
Total cost of production $/t			842			755 (768)

* Costs for H_2SO_4 recycling included in other cost positions

Source: [6.63]

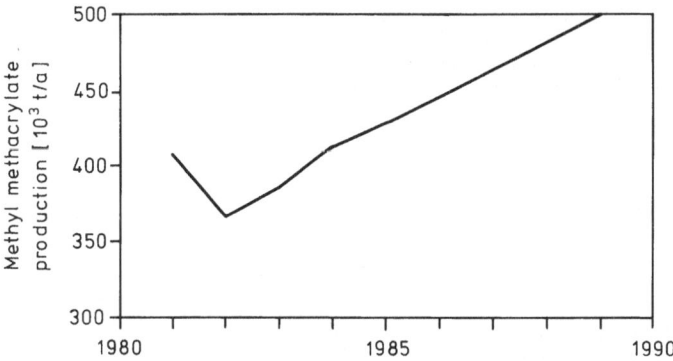

Fig. 6.15. MMA production in the USA 1981–1985, with a forecast up to 1989 [6.63]

Table 6.31. Consumption pattern for MMA in the USA 1985

Inland consumption	1000 t/a	%
Moulded sheets	60	14
Extruded sheets	85	20
Injection and other extrusion parts	80	19
MMA-modified acrylate resin varnishes	100	24
Impact-resistant plastic components	20	5
Higher methacrylates in the lubricant field	10	2
Styrene substitute in UP resins	10	2
Miscellaneous (leather finishing, paper coatings, adhesives, inks, floor polishes)	35	8
Total	400	94
Export surplus	25	6
Amount produced	425	100

gained favour, the growth rate of MMA after the slump in the USA in the early 1980s fell to 3 to 5%. Experts forecast an average yearly growth rate in the USA of at least 3.5% of MMA for the next few years, and an increase of the export surplus to 40,000 t/a. The production should thus reach 500,000 t/a in 1989 (see Fig. 6.15).

Producers of MMA, particularly Mitsubishi Rayon, are engaged in intensive research for new fields of application of MMA which could accelerate production expansion. The following uses may be mentioned [6.64]:

— carbonisation-resistant polymer concrete for swimming pools, sewage plant tanks, chemical plants, bridges, balconies, etc.
— CD-music and video discs
— anti-dazzle, reflecting television and computer monitor screens
— extremely scratch-resistant sheets
— highly transparent, unbreakable fibre light conductors (Eska fibres of Mitsubishi Rayon)

In the FRG, Röhm, Degussa and Resart-Ihm manufacture bead polymerisates, from which all three produce acrylate sheets (trade names: Plexiglas, Deglas, Resartglas).

Among new MMA specialities are:

— acrylonitrile-MMA (AMMA) co-polymers (e. g. Plexidur of Röhm)
— PVC-MMA-alloys (e. g. Kydex of Rohm and Haas)
— polymethacrylimide (PMI) hard foam rubber (Rohacell of Röhm)

The replacement of styrene by UV resistant MMA in UP resins and styrene copolymers is arousing increasing interest. Lamination with PMMA of sheets and moulded parts made from favourably priced ABS increases substantially resistance to weathering and UV-radiation and hence the service life. The increasing demand for high quality consumer goods made from PMMA (e. g. salt and pepper pots, bowls, mugs) is a favourable trend. Further, PMMA belongs to those materials with conspicuously good recycling properties since its pyrolysis enables the monomer to be recovered. This is an advantage from the ecological point of view. In its end uses PMMA competes with technical plastics, such as PBTP and ABS/SAN and especially PC. Thermoplastic materials for mass production, such as PS and PVC, and duroplastic UP resins are also competitors. Since all these competing products are cheaper and also easier to process, PMMA can succeed only through its superior properties for technical and consumer applications.

The facts that the TBA route suffers from difficulties neither of troublesome co-products nor of handling hydrocyanic acid, especially when the acid is supplied from outside, and also that it is cheaper and requires less capital investment makes preference for this route understandable, provided the necessary substitution and extension investments can be made. Even TBA producers can act as suppliers of MMA monomer by means of forward integration. The decision to favour the TBA-MMA process must be preceded, however, by considering that the ACH process can conveniently take, at correspondingly low contract prices, two surplus co-products as raw materials: acetone from the phenol syntheses according to the cumene process; and hydrocyanic acid from acrylonitrile syntheses using the Sohio process. In the FRG, hydrocyanic acid is a key base chemical of Degussa in the Wesseling factory, finding wide application for MMA, methionine, cyanuric chloride and cyanides. This ensures that HCN is produced at the lowest possible cost. Therefore HCN production costs according to the BMA process, taking into consideration the credit for the co-product, hydrogen, should determine the upper limit of the contract price in the FRG. Degussa obtains hydrocyanic acid from EC, Cologne-Worringen by tank wagon, also ACH from Chemie Linz. Röhm obtains ACH from DSM, Hoechst/Münchsmünster and Hüls. The other side of the picture is the fact that MMA processing requires considerable know-how. Additionally, the production of semi-finished goods and also the sale of sheets and profiles and their further processing to some extent, are in the hands of the present MMA producers. They resist any attempt of a new producer to establish additonal downstream activities in this market.

6.2.4.2 Poly-*i*-butene

High-molecular poly-*i*-butene (PIB) is a synthetic product with rubber-like properties, i. e. elastic under temporary stress (e. g. BASF Oppanol-B types of the series 100 to 200). PIB gives excellent service over a wide range of temperature from 50 to 100 °C. It has very low permeability to steam and gas, high resistance to weathering and aging, advantageous dielectric properties and is stable towards aggressive chemicals except concentrated nitric acid and halogens. It finds its principal application as an auxiliary in processing of plastic and rubber and in the form of waterproof sheeting.

Some producers add *i*-butene to polyethylene as a comonomer to improve the stability towards stress cracking. Stress cracks appear through the action of surface-active materials (e. g. cleaning agents and detergents) on parts which are under mechanical stress or during processing procedures (e. g. deep drawing) which create latent strain. It is an advantage that this effect can be attained with small amounts because of the high molecular weight of the PIB, without the hardness and rigidity of the product suffering as would be the case if propene were used as comonomer. Cross-linked polyethylene (XPE) is a competitor. The processing costs of this product are appreciably higher since the cross-linking has to be performed in a separate processing stage after shaping.

Addition of PIB to rubber mixtures is especially advantageous in the manufacture of tires because it increases the adhesion of the treads as well as improving stability towards weathering and aging and reducing permeability to air. PIB is a competitor here of the more expensive butyl rubber. Although this co-polymerisate of *i*-butene and isoprene has the advantage that it can be vulcanised to some extent, it is more sensitive to weathering because of the presence of the polyunsaturated isoprene. Only the halogenated butyl rubber, still more expensive, is sufficiently weather-resistant.

PIB plus fillers is processed as an independent product to sheeting for lining and sealing in buildings. In the FRG this sheeting is manufactured under the name of Rhepanol by the firm of Brass, Mannheim. It competes with sheeting from bitumen, polyethylene, soft-PVC, chlorinated polyethylene, chlorosulphonated polyethylene, and atactic polypropene.

The raw materials for PIB make sheeting from it more expensive than that from other plastics and especially that from bitumen. Rhepanol sheeting is thus used principally where there is a demand for its superior properties. Recommendations for the particular type of plastic sheeting to use in a given situation can be obtained from guidelines developed by a technical body established by the producers of plastic and rubber sheeting for sealing in roofing and building work (TAKK) at the Institute for Construction with Plastics (IBK) in Darmstadt in the FRG and also from standards committees and DIN-standards of the Association of German Engineers (VDI). Anyone who conflicts with these recommendations which are based on the present state of the technical art, by using a cheaper but technically less suitable product may have to pay compensation for subsequent building deficiencies.

6.2.4.3   Poly-1-butene

Because of its high molecular weight (compact crystal structure) isotactic poly-1-butene (PB) has, compared to other polyolefines, a distinctly higher stability towards stress cracking, greater abrasion resistance, a lower tendency to creep and good stability over a period, even at higher temperatures. PB can be subjected easily to extrusion, moulding, injection moulding and blow moulding at temperatures above 190 °C in the machines employed successfully for polyethylene [6.65]. On cooling from the thermoplastic state the phenomenon of polymorphism is shown, unusual for plastics. An unstable (soft) modification is first formed, crystallizing at room temperature during the next 7 to 10 days into the stable (hard) modification. This property may be an advantage or a nuisance for the particular processing procedure. The crystallization can be abbreviated with the help of special additives or the influence of pressure or tension [6.66] but this has acquired no practical importance [6.67].

On account of its markedly higher production costs compared to those of polyethylene, PB has had market success only in special fields of application since production began early in the 1970s (Table 6.32). In some fields its distinctive properties are acknowledged by the higher price acceptance. They include particular fields of pipe applications, notably for floor heating and, increasingly in the future, for hot water; also sheets and receptacles suitable for filling with hot material, e. g. bitumen bales and detergents [6.68].

Table 6.32. Cost of production for poly-1-butene

Product, purity:	isotactic Poly-1-butene (granulated)		
Raw material, specification:	Raffinate II, 50 wt. % 1-Butene		
Process:	Hüls PB		
Capacity:	12,000 t/a		
Capital investment	10^6 $	10^6 $	$/t · a
Basis	(IV 70)	(III 87)	
Fixed capital	5.5	11.4	950
Working capital		1.7	142
Total capital investment		13.1	1092
Production costs	Unit costs	Units/t	Costs $/t
Raw material costs			
Raffinate II	310 $/t	3.00 t	930
Catalyst and auxiliaries costs			160
Nitrogen	35 $/1000 m^3	300 m^3	11
Energy costs			
Electricity	7 ¢/kWh	950 kWh	67
Steam MP 20 bar	14 $/t	2.0 t	28
LP 3 bar	12 $/t	14.0 t	168
Cooling water	7 ¢/m^3	600 m^3	42
Cooling energy (−5 °C)	20 $/GJ	2.51 GJ	50
Material and energy costs			1456
Labour and overhead costs			
Operating labour	30000 $/my	10/shift	125
Operating supervision	45000 $/my	1	4
Plant overhead	100% excess		129
Capital related costs			
Fixed capital related costs	18%		171
Interest on working capital	9%		13
By-product credit			
Butanes 45.7 GJ/t	4 $/GJ	−1.7 t	−311
atactic Poly-1-butene	50 ¢/kg	−50 kg	−25
Total cost of production $/t			1562

Source: [6.69]

In principle, plastic pipes can compete with pipes from other material (copper, steel, concrete) only in the lower diameter sizes (up to a maximum of about 30 cm).

Compared to heating with radiators, under-floor heating yields a more comfortable room warmth at lower running costs. This is because the surface heating has a better thermal efficiency and hence manages with heat at lower temperature (preliminary heating temperatures of below 60 °C, and up to 40 °C in the heating pipes system) which permits the use of economical heat pumps and solar energy. The higher investment costs and the awkward controllability of the heat distribution are disadvantages.

If the energy supply is costly the higher investment costs for floor heating and heat pumps pay off. For the reason of rapidly rising energy prices at the end of the 1970s and the beginning of the 1980s, a peak demand was reached for especially this combination of floor heating and heat pumps. The demand was temporarily increased further by subsidies for energy-saving systems (see Table 6.33). At present it is more a consideration for comfort than cost which determines whether or not to install floor heating.

Practically only pipes of iron or copper were used up to the beginning of the 1970s for the — then still rare — floor heating. At first copper pipes were preferred because of their easier installation and the rapidly falling price of copper. Since then plastic pipes have become just as easy to install and have the advantage over metal pipes that their thermal conductivity is lower (PB: $\lambda = 0.22$ W/mK; Cu: $\lambda = 370$ W/mK). Plastic pipes thus distribute the heat more uniformly over the surface to be heated, or the heating medium can be circulated much more slowly. About two thirds of the European market for floor heating is in the FRG, in which about 90% are of plastic pipes and the remainder of copper or electric heating spirals. The competing plastic materials are PB, cross-linked polyethylene (XPE), or co-polymerised polypropene (PP-C), either as random co-polymer (PP-R, with 1 to 4% by wt. of ethylene manufactured by Hüls), or as ethylene-block co-polymer (PP-B, made by Hoechst). Table 6.33 shows the development of the demand for plastic pipes in this market.

Experts have criticised the accuracy of the data above, especially in connection with PP-C. It is believed that 30 million m PP-C pipes, in a total market for floor heating pipes of, at the most, 60 million m, were installed in the FRG in 1983 [6.72].

The sales slump in 1982 and 1983 was explained by corrosion damage to iron pipes, tanks and boilers due to probable diffusion of oxygen in the plastic pipes. Although most of the damage was the result of faulty installation, three alternatives have been defined in the proposals of standards for floor heating pipes (DIN 4726—4729) with which the technical aspect of the oxygen diffusion can be considered as solved:

— by installation of newly developed pipes from the same plastic materials but which are less permeable to oxygen
— by adding inhibitors to the circulating hot water which create a protective layer preventing corrosion (as in car radiator cooling water)
— by separating circulations in the plastic and steel pipes, or removing steel tanks or boilers from the circulation in the plastic pipes.

PB has maintained a largely constant fraction of the market of about 10%, but PP-C has lost some sales to XPE since several pipe manufacturers have received licenses for technical equipment for cross-linking polyethylene. The technology of

Table 6.33. Development of the market for plastic floor heating pipes in Western Europe 1980−1986 and forecast up to 1990 in 10^6 m/a installed [6.73]

FRG

	1980	1981	1982	1983	1984	1985	1986	1987	1988	1989	1990
PB	9.3	6.4	5.8	3.0	4.7	4.2	3.5	3.0	2.5	2.5	2.0
PP-C	58.0	29.4	13.9	13.2	10.2	8.4	9.0	9.0	8.0	7.0	7.0
XPE	48.7	55.2	38.3	36.0	35.6	30.4	32.0	36.5	42.0	47.0	51.0
Total	116.0	91.0	58.0	52.2	50.5	43.0	44.5	48.5	52.5	56.5	60.0

Switzerland

	1980	1981	1982	1983	1984	1985	1986	1987	1988	1989	1990
PB	1.2	2.1	3.0	2.5	2.2	2.0	2.0	2.2	2.3	2.4	2.5
PP-C	0.9	1.2	1.8	3.6	3.8	3.8	3.0	3.0	3.0	3.0	3.0
XPE	6.8	9.5	6.8	4.1	4.5	5.2	5.4	5.8	6.7	7.1	7.5
Total	8.9	12.8	11.6	10.2	10.5	11.0	10.4	11.0	12.0	12.5	13.0

France

	1980	1981	1982	1983	1984	1985	1986	1987	1988	1989	1990
PB	0.1	0.2	0.5	0.7	0.6	0.5	0.5	0.5	0.5	0.5	0.5
PP-C	3.3	3.7	2.6	1.6	1.0	1.0	1.0	1.0	1.0	1.0	1.0
XPE	3.4	4.5	4.7	4.9	5.4	6.0	7.0	7.0	7.5	8.0	8.5
Total	6.8	8.4	7.8	7.2	7.0	7.5	8.5	8.5	9.0	9.5	10.0

Austria

	1980	1981	1982	1983	1984	1985	1986	1987	1988	1989	1990
PB	1.1	1.9	1.0	0.8	1.5	1.6	1.7	1.8	1.8	1.8	1.9
PP-C	2.1	3.1	2.2	1.9	1.0	0.8	0.8	0.8	0.8	0.8	0.9
XPE	2.6	2.8	2.0	1.7	1.5	1.7	1.9	2.4	2.6	2.8	3.0
Total	5.8	7.8	5.2	4.4	4.0	4.1	4.4	5.0	5.2	5.4	5.7

Other countries (Benelux, Great Britain, Italy)

	1980	1981	1982	1983	1984	1985	1986	1987	1988	1989	1990
PB	0.6	0.7	0.7	1.0	1.0	1.1	1.2	1.3	1.4	1.5	1.6
PP-C	2.3	3.2	3.5	3.0	2.5	2.2	2.0	2.0	2.0	2.0	2.0
XPE	0.9	1.1	1.5	2.0	2.5	2.5	4.9	5.5	7.5	7.5	8.0
Total	3.8	5.0	5.7	6.0	6.0	5.8	8.1	8.8	10.9	11.0	11.6

W. Europe

	1980	1981	1982	1983	1984	1985	1986	1987	1988	1989	1990
PB	12.3	11.3	11.0	8.0	10.0	9.4	8.9	9.8	8.5	8.7	8.5
PP-C	66.6	40.6	24.0	23.3	18.5	16.2	15.8	15.8	14.8	13.8	13.8
XPE	62.4	73.1	53.3	48.7	49.5	45.8	51.2	57.2	66.3	72.4	78.0
Total	141.3	125.0	88.3	80.0	78.0	71.4	75.9	81.8	89.6	94.9	100.3

these three plastic products is discussed in more detail in the following pages in order to estimate their long-term specific competitive potentials.

PB is at a disadvantage compared to ethylene and propene because its monomer is more expensive (see Table 6.24, 6.25, and 6.26). These extra costs cannot be compensated for during the polymerisation stage because it is cheaper to mix peroxide with some polyethylene batches from large polymerisation plants (XPE) or to add ethylene as comonomer to some polypropene batches (PP-C) than to operate a small, and hence costly to run, unit for polymerising 1-butene. A possible opportunity for PB producers would be to make PB batchwise in new polypropene plants with flexible utilization.

On the other hand the manufacture of pipes from XPE is much more laborious than from PB or PP-C, on account of the complicated cross-linking procedures. Such procedures are as follows:

— admixture of special peroxide and treatment in the ram-extruder at pressures up to 10,000 bar (Engel process)
— grafting with organo-silicon and splitting off cross-linking radicals by the action of water (Sioplas process, Monosil process)
— irradiation with high-energy isotopes and formation of cross-linking radicals (process of cross-linking by radiation).

The total cost of a floor heating layout is made up of the costs of the material and the labour involved in installation. Since the three sorts of pipe can be largely laid identically (endless, flexible), the material costs determine primarily the price/m² of a floor heating. The material requirement can be derived from the basic standard for dimensions of floor heating pipes, in DIN 4725.

The pipe thickness is calculated from the data in this DIN (classified according to standardised outside diameter/static pressure) and the resistance to longtime stress (reference tensile strength ς_v) for the particular material, given in Fig. 6.16. The equation of this calculation is

$$s = \frac{S \cdot PN \cdot d_a}{20 \cdot \varsigma_v + S \cdot PN}$$

where s = wall thickness in mm, S = safety factor according to DIN, PN = static pressure in bar, according to DIN, d_a = outside diameter in mm according to DIN, ς_v = reference tensile strength in N/mm² according to DIN (value after 50 years for floor heating installations).

The weight of the pipe is obtained from the wall thickness and the outside diameter. It is evaluated per metre pipe for easier comparison according to

$$m = \pi \cdot \varrho \cdot (d_a - s) \cdot s$$

where m = weight of the pipe in g, ϱ = density of the material in g/cm³.

Examples of calculated values for a floor heating pipe of $d_a = 20$ mm are given here:

Plastic material	Wall thickness in mm	Pipe weight in g/m
PB	2.2	113
PP-C	2.7	133
XPE	1.7	93
copper pipe of $d_a = 18$ mm for comparison	1.0	457

The next highest standard wall thickness has to be chosen.

PP-C pipes can thus compete with PB only when they cost at least 15% less per kg. In reality XPE pipes cost about twice as much as PB pipes and two and a half times as much as PP-C pipes. Despite their considerably higher price in the FRG, XPE pipes

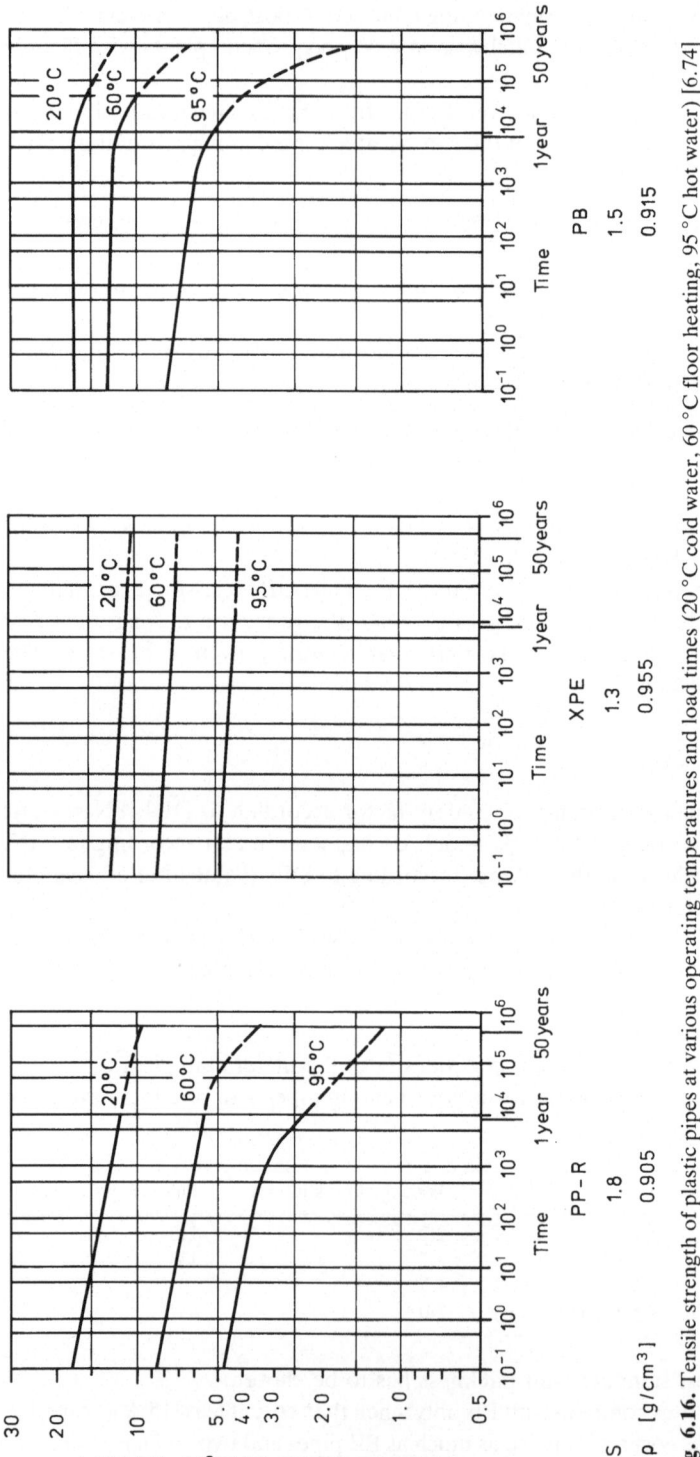

Fig. 6.16. Tensile strength of plastic pipes at various operating temperatures and load times (20 °C cold water, 60 °C floor heating, 95 °C hot water) [6.74]

have about 60% of the market for floor heating pipes. This results from the marketing policy of the suppliers who argue with empirical but unproven statements relating to the safety of their markedly more expensive product [6.76].

A thirty year guarantee is given by the manufacturers of all the plastic pipe materials mentioned, extending also to all secondary damage. The danger that the plastic floor heating pipes will lose their service durability can arise only after a period of use for far more than thirty years, according to the curves of resistance to continuous load (Fig. 6.16). The house owner alone has to bear the costs of this damage when the guaranty has expired and he is consequently ready to accept higher prices on hearing arguments about reliability.

Newer results indicate that the endurance properties of XPE pipes also suddenly slump after 10 years service life, contradicting the curves in Fig. 6.16 [6.75].

However, the test conditions of permanent stress are 10 bar and 60 °C, yet the operating conditions of floor heating pipes are only 1 to 2 bar and water temperatures of mostly 25 °C with only temporary rises to as high as 40 °C.

The supervision of the quality of plastic floor heating pipes is the responsibility, in the FRG, of the Gütegemeinschaft für Kunststoffrohre im Kunststoffrohrverein (GKV) e. V., Bonn. In addition, supervision agreements with testing institutes can be made on the basis of the pipe standards given here (in future DIN 4726):

PB   DIN 16 968/969
PP-C   DIN 8 077/8 078
XPE   DIN 16 892/16 893

As far as the marketing possibilities are concerned it must be kept in mind that even if only PB were used in the complete Western European market for floor heating pipes (about 80 million m/a) only about 9000 t/a PB would be needed.

Further sales expansion could come with the development of new fields of application, especially for hot water pipes. At present copper pipes still predominate here but with certain qualities of water (e. g. in alpine regions or the lower Rhine valley) they tend to show pitting corrosion. Even copper pipes "specially treated against pitting corrosion" (so-called Sanko pipes) are guaranteed by the manufacturers for only 5 years service life (including secondary damage). Their replacement by plastic pipes is proceeding, having started in the endangered areas. It is mainly by XPE pipes with screw clamps of red cast. The acceptance of plastic pipes has been supported by the so-called pipe-in-pipe system, introduced at the beginning of 1987. In the event of damage a defective plastic pressure pipe can then be easily replaced.

The standards commission is in the process of preparing DIN 1988. It aims to require hot water pipes to withstand hot water at 60 °C for 50 years, with 1000 hours temporarily at higher temperature (maximum of 85 °C in the given pressure stages). PP-C also fulfils these test conditions and needs no expensive screw clamps as does XPE and sometimes PB. Hence the chances of PP-C in the field of hot water pipes are conceded to be particularly good [6.67]. A maximum number of 400,000 new residences per year in the FRG and 50 m hot water pipes per residence yields a marketing volume of 20 million m of pipes. This corresponds to only 2300 t/a PB, based on the assumption that only this material is used in this field.

6.2.4.4   Processing of Other Polyolefines by Use of 1-Butene as Comonomer

Research on new, highly selective catalyst systems and development of new processes made possible the production of linear polyethylene of low density (LLDPE) at the end of the 1970s. Low-pressure processes and the use of larger proportions of suitable comonomers enabled products to be made with properties similar to those of the high-pressure LDPE, which has a disadvantage because of the high energy costs. 1-Butene has become quantitatively the most important comonomer in this LLDPE production, and for several reasons — it had been used successfully for a long time as comonomer in the preparation of HDPE; it is more cheaply obtained with high purity than any other α-olefine; and it is available in large amounts because it is required only in limited amounts in other fields, especially profitable ones.

A surge of new or modified plants for making LLDPE swept through the USA and they will be erected in many countries throughout the world in the near future (see Table 3.12). The question may be posed about the reasons for this development, from the point of view of both the process and the specific nature of the product.

Table 6.34 shows the cost estimation for LLDPE powder according to the gas phase fluid bed process of BP. The BP technology corresponds to the Unipol process of UCC, the most widely used one. The advantages over a conventional high pressure LDPE plant are as follows [6.77]:

- only half as much fixed capital investment is needed
- 75% less energy requirement (partly counterbalanced by loss of credit items for high pressure steam)
- greater flexibility of product properties (see Fig. 6.26)
- constant production amounts of all product qualities, in the high pressure process, the yield is cut by one half when high molecular film is produced instead of low molecular injection mould quality
- operation possible at low capacity utilization
- only one tenth of surface area is needed at the site of the unit
- lower maintenance costs
- construction time 8 to 12 months shorter
- fewer waste gas problems and explosion dangers

These momentous advantages have had the consequence that no new LDPE plants have been built in the western industrial countries for several years [6.71] and that old, shut down plants were being replaced by LLDPE plants as required.

In the FRG no replacement has followed the run-down of capacity by 380,000 t/a LDPE and 253,000 t/a HDPE in the years 1980—1983 [6.78]. The polyethylene capacities of Table 6.35 thus still exist.

This balanced LDPE/HDPE capacity enables, with the use of comonomers, the production of a wide range of qualities which can be surpassed by LLDPE gas phase processes only in peripheral domains (Fig. 6.17). Operating experience has demonstrated that the LLDPE gas phase processes can realise their cost advantages only when the plants run continuously without change of product type [6.79]. The losses occurring through more frequent product changes, due to reduced product qualities each time the large reactors are started up anew, can be balanced only by lengthening the production cycle and using suitable storage facilities which are expensive.

Table 6.34. Cost of production for LLDPE

Product, purity:	LLDPE (powder), 11 wt.% Comonomer		
Raw material, specification:	Ethylene, 1-Butene		
Process:	BP LLDPE		
Capacity:	120,000 t/a		
Capital investment	10^6 FF	10^6 \$	\$/t · a
Basis	(I 85)	(III 87)	
Fixed capital	370	67.8	565
Working capital		10.2	85
Total capital investment		78.0	650
Production costs	Unit costs	Units/t	Costs \$/t
Raw material costs			
Ethylene	410 \$/t	0.92 t	377
1-Butene	496 \$/t	0.10 t	50
	(see Table 6.24)		
Catalyst and auxiliaries costs			5
Nitrogen	35 \$/1000 m^3	50 m^3	2
Energy costs			
Electricity	7 ¢/kWh	165 kWh	12
Steam LP 3 bar	12 \$/t	0.1 t	1
Cooling water	7 ¢/m^3	4.5 m^3	0
Material and energy costs			447
Labour and overhead costs			
Operating labour	30000 \$/my	8/shift	10
Operating supervision	45000 \$/my	1	0
Plant overhead	100% excess		10
Capital related costs			
Fixed capital related costs	18%		102
Interest on working capital	9%		8
Total cost of production \$/t			577

Sources: [6.70; 6.71]

Table 6.35. Polyethylene producers and capacities in the FRG in 1985 (in 1000 t/a) [6.80]

Company		Works	LDPE	HDPE
BASF/Shell	ROW	Wesseling	400	215
Bayer/BP	EC	Cologne-Worringen	300	
Hoechst		Frankfurt		130
		Hürth-Knapsack		100
		Münchsmünster		80
	Ruhrchemie	Oberhausen	120*	70
Hüls		Marl		160
Total			820	755

* leased to Enichem until 1994

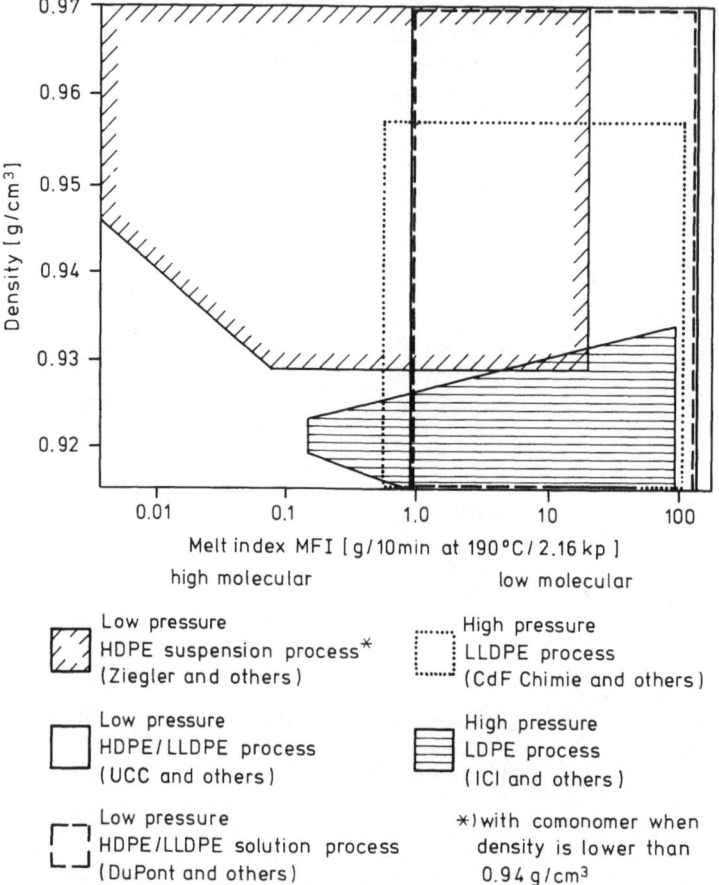

Fig. 6.17. Product properties of polyethylene processes [6.81]

Taking the melt index as a measure of the degree of polymerisation, LLDPE always has a higher average molecular weight than LDPE. This is due to its linear structure. This improves the mechanical properties of the products but the melt has then a higher viscosity. Since both its melting and softening temperatures are higher than those of LDPE, more energy is needed for processing LLDPE.

LDPE and LLDPE have closely similar main consumption patterns in Western Europe — 73 and 75%, respectively, of higher molecular film qualities, and 9% and 12%, respectively, of low molecular injection moulding, extrusion moulding and blow moulding qualities [6.81].

Hitherto LDPE has practically always been mixed with LLDPE for film production in Western Europe whereas it has largely been used alone in North America. This is due to the differing growth developments of the market structures for LDPE in the two regions. In Western Europe thick films for plastic bags and shrink films predominate, but on the American market thin sheets for packaging food and stretch films prevail [6.81].

LLDPE is a good material for stretch films but not suitable for shrink films [6.82]. LLDPE is particularly suitable for thin films because it can be extruded with negligible danger of tearing and has good hot sealing behaviour. It can be processed into films 30% thinner than those from LDPE but of comparable resistance to tearing [6.83]. LLDPE films can be drawn out to a thickness of 7 µm, LDPE films on the other hand to no thinner than 20 µm [6.84]. For this, however, it is necessary to adapt the film blowing equipment, optimised for high efficiency with LDPE. This requires investment outlay and the processors are insufficiently encouraged by the saving in material. They have to adjust the extruder to the higher shearing viscosity of the melt, and the blowing equipment to the lower extension viscosity of the film [6.85]. A strategy of low price was chosen in the USA for introducing LLDPE on the market which caused the processors to increase investment. Up to now there have been no price advantages for LLDPE over LDPE in Western Europe (Fig. 6.18).

The packaging industry is enjoying many innovations which provide fertile ground for films and replacement chances via development of packaging or wrapping machines. This makes it difficult to forecast quantities in the LLDPE market. The thin LLDPE films compete not only with LDPE and HDPE but also with PVC, PVDC, VCVDC (Saran), OPP (biaxially oriented polypropene), Cellophane and composite films of these plastics with Cellophane. The OPP films have the largest growth rate at present, mainly at the expense of Cellophane. It is highly transparent and stable. Therefore its production has been expanded [6.86].

Approximately 400,000 t/a LLDPE, mostly added to LDPE, were used in 1985 in the Western European film sector. Esso, as importer from Saudi Arabia, counts on

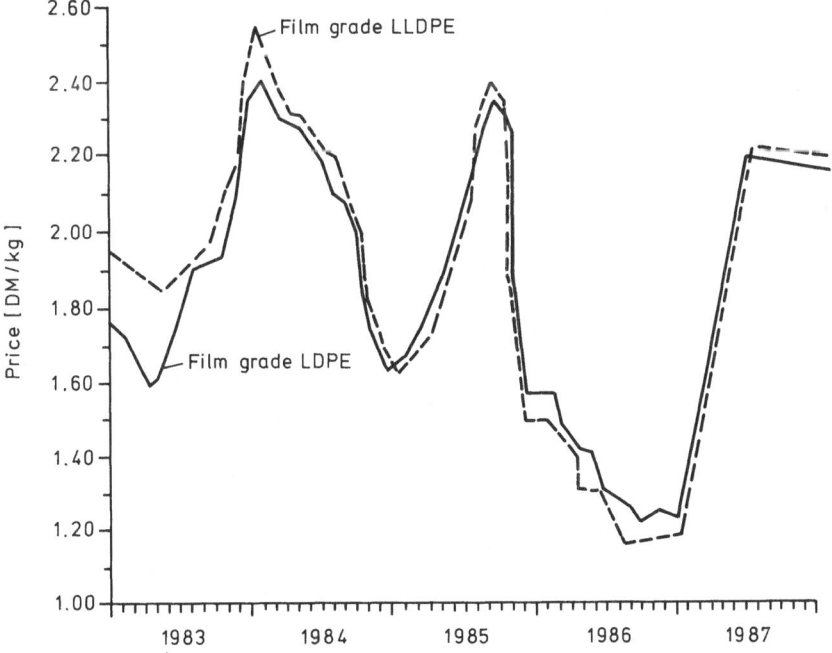

Fig. 6.18. Market prices for LDPE and LLDPE films in Western Europe 1983−1987 [6.87]

Table 6.36. Fields of application of polyethylene injection moulded parts, related to properties of the product

Melt index MFI in g/10 min at 190 °C and 2.16 kp	Density in g/cm³				
	0.92	0.93	0.94	0.95	0.96
>25-15	flowing most easily, low duty mass products	flowing easily, parts of large surface area, good dimensional stability, good gloss	flowing easily, shock-proof parts, no special requirements for rigidity	flowing easily, good dimensional stability, household articles with difficult injection techniques	flowing easily, very hard and rigid, bowls, sieves, crockery, transport boxes, protective helmets
15-5	more stable articles; little surface gloss	parts of little internal tension and glossy surface	shock-proof, low tendency to stress cracking, high duty technical parts	easily workable, highly shock-proof, screw caps, closing devices, technical parts	shock-proof, high dimensional stability, parts subject to high mechanical stress, e. g. trash cans, seats
~1.5	very good mechanical stability and resistance to stress cracking		good long term stability, little tendency to stress cracking, high duty closing devices for special purposes	stable against stress corrosion, good surface, technical parts subject to high mechanical stress	
<1				high-molecular, mostly highly stabilised, fittings resisting pressure, pipe elbows and fittings	

LLDPE, with yearly growth rates of over 20% in this field, passing the 1 million t/a figure in Western Europe by 1990 [6.88]. BP, as a European producer of LLDPE, assumes a 12% yearly growth rate, which would lead to 700,000 t/a by 1990 [6.89].

LLDPE has attained about 20% of the Western European LDPE injection moulding market, the total of which is about 400,000 t/a, on account of its better tensile strength, impact resistance, resistance to stress cracking and stability at low and high temperatures compared to LDPE. The LLDPE market share in the USA is about 70% and in Canada, almost 100% [6.90]. The best replacement chances for LLDPE are in the density range of 0.94 g/cm^3, covered by LDPE/HDPE blends and copolymer-HDPE for injection moulded parts (Table 6.36).

HDPE parts have a dull surface whereas LLDPE fulfils the desire for a shinier surface and permits the processing cycles to be shortened at the same time. In the low molecular range of smaller density, LDPE shows the best processing properties in the manufacture of mass products, most of which have no high quality demands. For mass products goods of higher quality, polypropene is preferred since it has much superior properties to LLDPE.

Other substitution opportunities for LLDPE are for injection and rotationally moulded large parts with a low tendency to stress cracking which are made from cross-linked HDPE [6.85]. In contrast to LDPE, LLDPE is less suitable for blow-formation of hollow bodies and quite unsuitable for extrusion coatings [6.81; 6.92].

According to estimates of Esso, the Western European market for injection moulding will be penetrated up to 50% with LLDPE by 1990 [6.93]. Table 6.37 summarizes the consumption pattern for LLDPE in Western Europe in 1986. Figure 6.19 shows the increase in consumption of LLDPE in Western Europe between 1981 and 1986, and also gives a forecast for 1990.

Hitherto all the LLDPE needed in the FRG has been imported. Brenntag, a partner of Veba, has been appointed by Saudi Basic Industries (Sabic), Riad, as the sole marketer of Saudi LLDPE in the FRG [6.95]. Because of the generally small profit margin of mass thermoplastics, the producers in the FRG were, for a long time, not willing to motivate the plastics processors to abandon LDPE by means of a low

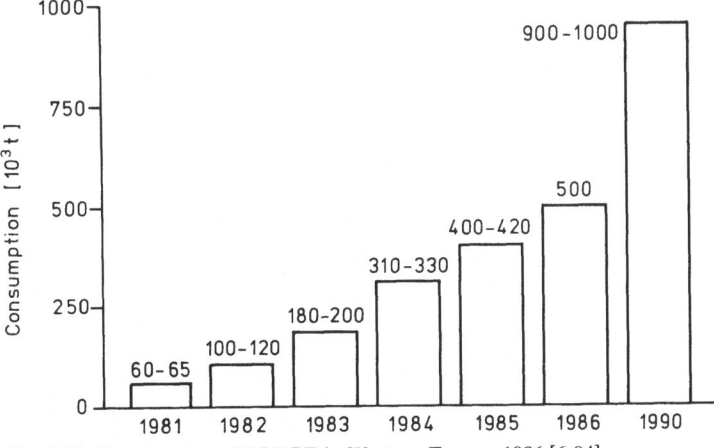

Fig. 6.19. Consumption of LLDPE in Western Europe 1986 [6.94]

Table 6.37. Consumption pattern of LLDPE (including blends) according to uses in Western Europe 1986 [6.94]

Field of application	Consumption (1000 t)
Films	410 to 420
of which:	
Heavy goods and sacks for waste	130
Stretch films	90
Shrink films	50
Shopping bags	30
Food packaging	25
Agricultural films	15
Rotationally formed parts	35
Wire and cable sheathing	15
Injection moulded parts	20 to 25
Pipes and pipelines	10 to 15
Miscellaneous*	10
Total	500 to 520

* including blow moulded parts, nets, foam plastics, masterbatches

price strategy for LLDPE. Instead, they have been waiting until the substitution, slower and determined by quality alone, has created a market of sufficient size, permitting a price increase justified by the quality of LLDPE and enabling an LLDPE producer to enter the field [6.96].

In order to be able to evade a price war of LLDPE standard qualities with 1-butene comonomer, market tests of LLDPE specialities with other comonomers, some new, and new brand names are being carried out. These comonomers involve longer chain α-olefines — up to C$_8$-olefines at the most because of problems of odour and handling which are encountered then. Among these, especially good prospects in the LLDPE film market are imputed to 4-methyl-1-pentene which imparts the products excellent optical and mechanical properties [6.97]. This comonomer derived from propene is produced exclusively by BP and is being used in their new LLDPE plant in Lavera. Other potential producers are the joint venture partner Neste Oy and the licensees USI, Chevron, and Mitsui [6.97]. UCC also has announced the commercialization of a new very low density polyethylene (VLDPE) generation with new comonomers [6.98]. The density of VLDPE is between 0.900 and 0.915 g/cm^3 and the product is suitable for partial substitution of ethylene-vinyl acetate copolymers [6.98].

Dow use only very small amounts of 1-octene as comonomer in their LLDPE types (Dowlex) and obtain products which impart better mechanical properties to the processed materials than LLDPE types containing higher proportions of 1-butene. Since it has a narrower distribution of molecular weight during processing the viscosity falls less markedly with the shearing gradient [6.100].

New investigations of the degradation of LDPE, HDPE and LLDPE by irradiation are interesting from the ecological viewpoint. This leads to hydrogen and other volatile products, mainly butanes and butenes, which can be reintroduced at the raffinate I stage and thus could contribute to recycling within C$_4$-chemistry [6.101].

7 Summary

The study of the competitiveness of C_4-hydrocarbons is subdivided into the presentation of the technologies for producing and processing them, the evaluation of the economics of these technologies, and the cost-benefit analysis of the secondary products. The result can be summarized in statements about the conditions of competition in the production process of the primary and raw materials, about the conversion alternatives into secondary products, and the assessment of the competitiveness of the industrial and speciality chemicals and the articles which can be made from them.

Of the C_4-hydrocarbons, exploitation of the i- and n-butenes is gaining more and more importance after butadiene has already been finding extensive use. Both monoolefines are yielded as co-products in refinery and chemical processes. In the refinery field their production amounts depend on the extent of gasoline production in conversion units, mainly catalytic crackers. In the chemical sphere they are produced along with butadiene in steam crackers as by-products together with ethylene and propene.

Because of their higher olefine content, the steam cracker C_4-hydrocarbons are of prime interest for use in the chemical industry. They are yielded in relatively large quantities in W. Europe and Japan, in contrast to N. America, and scarcely at all in the Arabian countries. This is due to differences in the feedstocks used for cracking. In N. America and the Arabian countries, and to some extent also in the U. K. and Norway, the cheaper ethane is preferred whereas the steam crackers in the other European countries and Japan are fed mostly with naphtha and still heavier materials. A change is on the way in the supply of petrochemical basic materials to the chemical industry. Ethane is running out both in the USA and the North Sea fields. Large amounts of ethane have been remaining mostly unused and available in the Arabian countries. Ethane is difficult to transport over large distances and therefore it is reasonable to carry out more cracking of ethane and production of ethylene there on the spot. Further, it is cheaper to convert the ethylene gas to secondary products at the production site rather than transport it with gas tankers to the industrialized nations for further treatment. The mass production of chemical products based on ethylene is thus being partly displaced to the Arabian countries.

In order to counter the risk of being put under pressure by the oil countries in connection with ethylene and even its secondary products, the large chemical concerns of the industrialized nations have to create flexible alternative petrochemical supplies. There is thus a tendency for the naphtha steam crackers here to aim not at maximum yields of ethylene but of propene. This increases, at the same time, the yield of C_4-olefines and ensures its expanded utilization.

Butadiene has always yielded, per unit quantity, the highest return of all the products from steam cracking. This is due to the comparatively high profit from the production and sale of the butadiene derivatives. Only since they have been in increasing demand for making MTBE or LLDPE have butenes gained transfer prices higher than those corresponding only to their heating value.

As a result of the MTBE boom of recent years the chemical industry is now competing with the mineral oil industry for i-butene. This competition will become fierce when leaded gasoline has been completely replaced by the unleaded product and if crude oil prices are high.

As long as the alternative prevails — that of producing MTBE from butene-containing refinery gases (refinery B-B) which is flared in most regions — the transfer price for i-butene-containing raffinate I will be derived from the forward-backward calculation of the MTBE production according to refinery B-B/raffinate I MTBE processes. Should there also be a shortage of i-butene from refinery B-B, as for example created by the production of alkylate gasoline in N. America, the supply of i-butene can be increased by isomerisation and dehydrogenation, or by isomerisation and oxidation to TBA (Arco propene oxide process) and TBA decomposition, all starting from n-butane derived from LPG (wet natural gas or refinery gas).

It is evident that the production of butenes by dehydrogenation of butane is uneconomic, even at the lowest raw material prices (e. g. in Saudi Arabia). The same applies to dehydrogenation to butadiene. On the other hand, the oxidation route is competitive in regions where the petrochemical structure permits the utilization of the co-product, propene oxide. The utilization of this C_4-technology guarantees that an appreciable price increase of i-butene, due to shortages, can be largely excluded in the long run. As a result of technical-economic conditions, the Arabian oil countries have only limited possibilities of building up a C_4-chemistry.

The LLDPE boom will similarly lead, with some time delay, to a shortage of n-butenes, especially 1-butene from the steam cracker C_4-cut. The raw material basis for 1-butene can scarcely be extended through using refinery B-B by molecular sieve extraction to obtain 1-butene, for reasons of cost and quality and, in N. America, also because of the large demand for n-butene to make alkylate gasoline.

It is thus reasonable, in this study, to set the transfer price for the 1-butene-containing raffinate II through forward-backward calculation of the production costs for 1-butene with the process with the lowest cost and, at the same time, not subject to raw material supply restrictions and gives a product of equal quality. The IFP Alphabutol process based on ethylene meets all these requirements. This gives credits per unit quantity for raffinate II which are still higher than those of the above-mentioned raffinate I and increase or decrease proportionally to the price of ethylene.

The production costs in C_4-chemistry suffer from the energy- and capital-demanding isolation of some pure C_4-components and from the economic necessity of utilizing to the maximum each of these components yielded in fixed ratios as co-products. Consequently C_4-technologies can be realised on the large scale only if it is guaranteed that sufficient amounts of these individual components can be converted into valuable end products.

The newer C_4-chemical processes are characterized by a high technological standard. Numerous selective direct syntheses, e. g. oligomerisation of n-butene, permit the use of mixtures of C_4-hydrocarbons rather than of pure components. Some processes are of high raw material and product flexibility, e. g. LLDPE procedures. Further, environmentally favourable processes have been available for some years, e. g. the ion exchanger procedure for obtaining the C_4-alcohols TBA and SBA in place of the former conversions using sulfuric acid with problems of residue disposal. Most of these technologies are available to all producers via licenses.

The successful application of these technologies depends ultimately on the degree of competitive success attained by their end products. The basis for this is their competitive potential regarding specific cost, quantity and quality. These are analysed, taking as examples the motor fuel and plastic sectors, because between these two exists the main competition for the butenes.

The studies of competition of motor fuel components have shown that, compared to other high-octane components which are available from refinery processes, the oxygenates derived from C_4-feedstocks (TBA/methanol, MTBE) have a higher product efficiency and, when the crude oil price exceeds a certain value, yield higher profits because of comparatively low production costs. This is only up to an optimum degree of blending with the gasoline. These oxygenates have a good chance, under favourable conditions, of filling the gap left by renouncement, for ecological reasons, of the lead alkyls and benzene; and they can, up to the legally limited quantity, attain additional market success at the expense of other gasoline components.

Two polymers can be produced in the plastics sector using i-butene and two using 1-butene (polymethyl methacrylate and poly-i-butene, and poly-1-butene and HDPE-comonomer/LLDPE, respectively).

The classical acetone/hydrocyanic acid process (ACH process) competes in the production of the monomeric methyl methacrylate. This route is still the most widely employed. Only in Japan has the i-butene/TBA route been adopted. The TBA route avoids disagreeable residues and problems connected with handling hydrocyanic acid (especially transport risks if it is not available on the site). It is also somewhat cheaper and needs less investment. Preference for it in necessary extension of substitute investment would be understandable. However, it must be noted that the ACH process is especially suitable for utilizing two co-products, presently available in surplus amounts (acetone from phenol synthesis, hydrocyanic acid from synthesis of acrylonitrile). If these products were not disposed of in this way, they would probably be obtainable at still more favourable contract prices. The production costs for polymethyl methacrylate are considerably higher than those for the plastic materials which compete with it. Its processing is also more complicated and therefore more expensive. Despite the lower prices of plastic products with equivalent application, its outstanding engineering properties enable it to take part in replacement competition quite efficiently.

Poly-i-butene is the oldest plastic product within C_4-chemistry and has been historically established in its various fields of application. It has competitors for neither the raw material nor the process, but only from other products. It has superior quality to other plastics with rubber-like properties, e.g. the cheaper polybutene which is obtained from a butene mixture. It is, however, inferior to the more expensive butyl rubber in special fields of employment, particularly to the halogenated form. Poly-i-butene is processed with filler to make waterproof sheeting for buildings. These are more expensive than those from other plastics or bitumen but find use where their superior properties are in demand or where building standards require their use.

Poly-1-butene is also without competitors for raw materials and processes but has to compete with other products. Its production costs are high so that it has only special fields of application in which its properties are acknowledged as value for money. These fields include the special use for pipes, notably for floor heating and, increasingly, hot water. Analyses of competitiveness yield no great advantage in properties over its rivals, cross-linked polyethylene and copolymerised polypropene. It also has no cost advantages over polypropene as a result of the relatively high costs of obtaining the monomer and of the polymerisation process. The total market in the above-mentioned field of pipe usage will require less than 10,000 t/a of plastic in the FRG in the near future.

Other polyolefines have been undergoing refinement for a long period by using small amounts of 1-butene as comonomer. A sudden jump in the demand for 1-butene was initiated by a process innovation for making linear polyethylene of low density (LLDPE). The low pressure gas phase process using larger comonomer fractions has been producing an LLPDE since the end of the 1970s having a range of properties similar to those of the high pressure LDPE which is more expensive to make because of higher energy costs.

The replacement chances for LLDPE depend on its market prices and on the structure of the film and injection moulding markets which vary greatly internationally. The considerable replacement success enjoyed by LLDPE in N. America is not expected in Europe because the polyethylene producers here follow a different strategy. Recently other comonomer specialities for LLDPE have been gaining in importance some of which can be synthesized from n-butenes.

The LLDPE required in the FRG has so far been imported and pure 1-butene as comonomer exported in appreciable quantities. The large scale plant for LLDPE in the FRG is to be built by EC in Cologne-Worringen.

Maleic anhydride (MA) and fumaric acid (FA) are intermediates for producing unsaturated polyesters. They can be made from benzene, n-butane or raffinate II. Because of its increased price through scarcity (due to LLDPE) raffinate II is losing its competitive ability. n-Butane can compete as a raw material for making MA only when its price is 100–125 \$/t below that of benzene. The world market price for benzene was so high in the USA before it was forbidden in gasoline that this price condition was comfortably fulfilled. The prohibition of benzene in gasoline is now imminent in W. Europe. Price-depressing surpluses of benzene are thus expected to increase in the world. They favour the benzene route decisively.

The butane basis would be substantially cheaper for the preparation of FA but even a drastic reduction of the production costs of FA would lower only slightly the material costs of the products. Furthermore, the techniques of the FA-processing are more complicated, creating higher costs, and the quality of the product is not so conspicuously improved that it would be possible to advance into the field of utilization of, notably, the epoxy resins. Consequently the big polyester processors hardly see any advantages in changing from MA to FA polyesters.

The C_4-secondary products used in both the plastic and fuel sectors are specialities. They possess this status because of the large utilization benefit of their end products. It is also typical that they are made and processed only by certain chemical companies which have created the infrastructure for the chemical utilization and possess the technical know-how for processing and using this group of products.

The C_4-secondary products in highest demand are those with the character of auxiliary and performance chemicals. The main products to which they are added (gasolinc, LLDPE) are mass produced and so, although specialities, these secondary products enjoy sales of the order of magnitude of mass products.

The best future opportunities for i-butene, in relation to all the C_4-base chemicals, appears to be in the motor fuel sector. There is a huge demand for components of high octane number and this is still increasing. It is far greater than what the whole chemical industry consumes of ethylene and propene together, i.e. the most important petrochemical raw materials. The consumption of i-butene in the plastics branch for making poly-i-butene and even MMA is, relatively, so trifling that these amounts are permanently available, particularly because these are special polymers for which not the price but their distinctive properties primarily determine the market outlets.

1-Butene from steam crackers is virtually the only n-butene which is of interest to the chemical industry. The demand of the motor fuel sector for n-butene is growing more slowly and is essentially limited to alkylate gasoline, so that refinery B-B suffices as the raw material basis. There is no procurement competition between the fuel and the plastic sectors because 2-butene provides higher octane qualities than 1-butene in the production of alkylate gasoline.

However, the steam cracker n-butenes are directly dependent on the mineral oil branch because naphtha is the raw material preferred for cracking. The naphtha specification corresponds to a low-value gasoline. Nevertheless the mineral oil industry has been trying for a long time to sell it to chemical firms at relatively high prices in order to participate in the high profits from its further processing in the chemical industry.

Producers who acquire naphtha for cracking to ethylene and propene and do not make use of the C_4-cut can evade this oligopolistic price policy of the mineral oil industry only to a limited extent by being flexible and employing other feedstocks for their crackers. On the other hand chemical firms with an understanding of how to utilize the C_4-cut are better placed for competition than are the mere users of ethylene and propene as a result of the profit contributions yielded by the C_4-components. The mineral oil concerns have no influence on the conditions of quantity and price of the C_4-cut producers within the chemical industry.

References

Chapter 1

1.1 Clementi A et al. (1979) Hydrocarbon Proc. 58, No 12: 109
1.2 Weitz H M, Hartig J (1975) Butadien, Ullmann (4), vol 16, p 13, Verlag Chemie, Weinheim
1.3 Mueller H J, Loeser E (1986) Butadiene, Ullmann (5), vol 4, p 440, Verlag Chemie, Weinheim
1.4 Obenaus F, Droste W, Neumeister J (1986) Butenes, Ullmann (5), vol 4, p 488, Verlag Chemie, Weinheim

Chapter 2

2.1 Obenaus F, Droste W, Neumeister J (1986) Butenes, Ullmann (5), vol 4, p 487, Verlag Chemie, Weinheim
2.2 Pujado P R, Vora B V (1983) Chem. Economy & Eng. Rev. 15: 16
2.3 Kuper H (1982) Erdoel u. Kohle 35: 119
2.4 Keim W, Behr A, Schmitt G (1986) Grundlagen der industriellen Chemie, p 65, Sauerlaender, Frankfurt, Berlin, Muenchen
2.5 Zahlen aus der Mineraloelwirtschaft-Ausgabe 1987, p 22, Deutsche BP, Hamburg 1987
2.6 Hamann W (1982) Erdoel u. Kohle 35: 342
2.7 Haun R R (1984) Chem. Eng. Prog. 80, No 1: 17
2.8 Price data ex Petroleum Economist
2.9 Rohoelimporte, Erdoelinformationsdienst (1987) No 5, p 11
2.10 Klamann D (1981) Petrochemie, Lecture at Technische Universitaet Berlin
2.11 Asinger F (1971) Die Petrolchemische Industrie, vol 1, p 7, Akademie-Verlag, Berlin
2.12 Little A D (1983) Auswirkungen des Wegfalls von Bleiverbindungszusaetzen und der Einstellung des Benzolgehalts von Benzin auf die deutschen Raffinieren, Part II, p 27, Umweltbundesamt, Berlin
2.13 Wetter F (1974) Erdoel u. Kohle 27: 77
2.14 Butadiene is swinging on a pendulum, Chem. Week, p 26, 23. 9. 1987
2.15 Butadiene Annual 1984, De Witt & Comp, Inc., Houston, TX, partly published in: Mueller H J, Loeser E (1986) Butadiene, Ullmann (5), vol 4, p 441, Verlag Chemie, Weinheim
2.16 ECN Bulk chemical prices
2.17 Freitas E R, Gum C R (1979) Chem. Eng. Prog. 75, No 1: 73
2.18 McLain S J, Schrock R R (1978) J. Am. Chem. Soc. 100: 1315
2.19 Butene-1, Petrochemical Process Handbook (1987) Hydrocarbon Proc. 66, No 11: 68
2.20 Sabic unit's butene-1 plant starts up at Jubail, Oil & Gas J. 85: 20, 26. 10. 1987
2.21 A renaissance in C4 chemistry, Chem. Week, p 36, 9. 9. 1987
2.22 Logan R S, Banks R L (1968) Hydrocarbon Proc. 47, No 6: 135
2.23 Johnson P H (1967) Hydrocarbon Proc. 46, No 4: 149
2.24 Hahn H-D, Daembkes G, Rupprich N (1986) Butanols, Ullmann (5), vol 4, p 465, Verlag Chemie, Weinheim
2.25 Weisz P B (1965), Erdoel u. Kohle 18: 525
2.26 Homann M, Schulze J (1983) Erdoel u. Kohle 36: 224
2.27 Mueller H J, Loeser E (1986) Butadiene, Ullmann (5), vol 4, p 435, Verlag Chemie, Weinheim
2.28 Chang C D et al. (1978) Ind. Eng. Chem. Proc. Res. Dev. 17: 255
2.29 Haggin J, Chem. & Eng. News, p 22, 22. 7. 1987

2.30   Asinger F (1986) Methanol – Chemie- und Energierohstoff, p 202, Springer, Berlin Heidelberg
        New York
2.31   Kaeding W W, Butter S (1980) J. Catalysis 61: 155
2.32   Asinger F (1986) Methanol – Chemie- und Energierohstoff, p 191, Springer, Berlin Heidelberg
        New York
2.33   Koenig L, Gaube J (1983) Chem.-Ing.-Tech. 55: 14
2.34   Afschar, A S, Biebl H, Schaller K (1987) Chem.-Ing.-Tech. 59: 509
2.35   Mueller H J, Loeser E (1986) Butadiene, Ullmann (5), vol 4, p 434, Verlag Chemie, Weinheim
2.36   Bach H, Gaertner R, Cornils B (1986) Butanals, Ullmann (5), vol 4, p 448, Verlag Chemie,
        Weinheim
2.37   Cornils B, Mullen A (1980) Hydrocarbon Proc. 59, No 11: 93
2.38   Hahn H-D, Daembkes G, Rupprich N (1986) Butanols, Ullmann (5), vol 4, p 463, Verlag
        Chemie, Weinheim
2.39   Hahn H-D, Daembkes G, Rupprich N (1986) Butanols, Ullmann (5), vol 4, p 468, Verlag
        Chemie, Weinheim
2.40   Asinger F (1986) Methanol — Chemie- und Energierohstoff, p 120, Springer, Berlin Heidel-
        berg New York
2.41   Eni/Haldor Topsoe unveil MAS technology, Europ. Chem. News 44, p 19, 10. 11. 1986
2.42   Falbe J, Weber J (1975) Butyraldehyde, Ullmann (4), vol 9, p 44, Verlag Chemie, Weinheim
2.43   Hahn H-D, Daembkes G, Rupprich N (1986) Butanols, Ullmann (5), vol 4, p 467, Verlag
        Chemie, Weinheim
2.44   Statistisches Bundesamt Fachserie 4, Reihe 3.1, Meldenr. 423116
2.45   Oxirane, Informations Chimie (1979), No 188, p 175
2.46   GTBA, Users Technical Manual 1982, p 2, Arco Chemical Europe Inc.
2.47   Fliege W, Voges D, Steffan G (1975) Butan-, Buten- und Butindiole, Ullmann (4), vol 9, p 22,
        Verlag Chemie, Weinheim
2.48   Davy & Sohio link in gas-to-thermoplastics process, Europ. Chem. News 44, p 20, 17. 11. 1986
2.49   Tamura M, Kumano S (1980) Chem. Economy & Eng. Rev. 12, No 9: 32

Chapter 3

3.1   Asinger F (1971) Die Petrolchemische Industrie, vol 1, p 380, Akademie-Verlag, Berlin
3.2   Buehler J H et al., Report on explosion at Union Carbide's Texas City butadiene refining unit,
        Chcm. Eng. 77, p 77, 7. 9. 1970
3.3   Kroenig W (1968) Erdoel u. Kohle 21: 140
3.4   Lauer H (1983) Erdoel u. Kohle 36: 249
3.5   Derrien M et al. (1979) Hydrocarbon Proc. 58, No 5: 175
3.6   Nierlich F, Obenaus F (1986) Erdoel u. Kohle 39: 73
3.7   Mueller H J, Loeser E (1986) Butadiene, Ullmann (5), vol 4, p 439, Verlag Chemie,
        Weinheim
3.8   Wagner U, Weitz H M (1970) Ind. & Eng. Chem. 62, No 4: 43
3.9   Dalen J. D. van, Zomerdijk J C, Shell uses acetonitrile for butadiene extraction, Europ.
        Chem. News 24, p 52, 30. 9. 1966
3.10   Peter W D, Rogers R S (1968) Hydrocarbon Proc. 47, No 11: 131
3.11   Takao S (1966) Hydrocarbon Proc. 45, No 11: 131
3.12   Coogler W W, Butadiene recovery process employs new solvent system, Chem. Eng. 74, p 99,
        31. 7. 1967
3.13   Mueller H J, Loeser E (1986) Butadiene, Ullmann (5), vol 4, p 437, Verlag Chemie,
        Weinheim
3.14   Baumann G P, Smith M R (1954) Petrol. Refiner 33, No 5: 156
3.15   Valet A M et al. (1963) Erdoel u. Kohle 16: 100
3.16   Kroeper H, Schloemer K, Weitz H M (1969) Erdoel u. Kohle 22: 605
3.17   Schulze J, Homann M (1986) Untersuchung der Vermeidungs- und Verwertungsmoeglich-
        keiten von schwefelhaltigen Rueckstaenden der anorganischen und organischen Zwischen-
        produktenchemie, Research report no. 103 01 343, Texte 16/86, p 274, Umweltbundesamt,
        Berlin
3.18   Wiesner J (1978) Umweltfreundlichere Produktionsverfahren in der chemischen Technik,
        Ullmann (4), vol 6, p 155, Verlag Chemie, Weinheim

3.19   Persak R A et al. (1978) Chem. Economy & Eng. Rev. 10, No 7: 25

3.20   Adler M S, Johnson D R (1979) Chem. Eng. Prog. 75, No 1: 77

3.21   Swodenk W (1984) Chem.-Ing.-Tech. 56: 1

3.22   Nierlich, F (1987) Erdoel Erdgas Kohle 103: 486

3.23   Production of *tert*-butanol and isobutylene, Chem. Eng. 90, p 60, 12. 12. 1983

3.24   Miranda M (1987) Hydrocarbon Proc. 66, No 8: 51

3.25   Direkthydratisierung zu *sec*-Butanol (1983) Europa Chemie, No 18: 326

3.26   *sec*-Butanol, Petrochemical Process Handbook (1987) Hydrocarbon Proc. 66, No 11: 67

3.27   Clementi A et al. (1979) Hydrocarbon Proc. 58, No 12: 109

3.28   Convers A, Juguin B, Torck B (1981) Hydrocarbon Proc. 60, No 3: 95

3.29   A renaissance in C4 chemistry, Chem. Week, p 36, 9. 9. 1987

3.30   Obenaus F, Droste W, Neumeister J (1986) Butenes, Ullmann (5), vol 4, p 489, Verlag Chemie, Weinheim

3.31   C4 separation by extractive distillation, Chem. Week, p 24, 26. 8. 1987

3.32   Marceglia G, Oriani G (1982) Chem. Economy & Eng. Rev. 14, No. 6: 35

3.33   Butamer, Refining Process Handbook (1980) Hydrocarbon Proc. 59, No. 9: 168

3.34   C4 isomerisation, Refining Process Handbook (1980) Hydrocarbon Proc. 59, No 9: 169

3.35   King R W (1966) Hydrocarbon Proc. 45, No. 11: 189

3.36   Craig R G, Duffalo J M (1979) Chem. Eng. Prog. 75, No 2: 62

3.37   Reidel J C, Making butadiene — Part 3: Butane Dehydrogenation today employs fixed-bed methods, Oil & Gas J. 55, p 114, 9. 12. 1957

3.38   Berg R C et al., Catalytic LPG dehydrogenation fits in '80's outlook, Oil & Gas J. 78, p 191, 10. 11. 1980

3.39   Craig R G, White E A (1980) Hydrocarbon Proc. 59, No 12: 111

3.40   Catofin, Petrochemical Process Handbook (1985) Hydrocarbon Proc. 64, No. 11: 149

3.41   Remirez R, No-lead gas turns into a world boom for MTBE, Chem. Eng. 94, p 19, 25. 5. 1987

3.42   Chowdhury J, Process trio selectively produce olefins from LPG, Chem. Eng. 91, p 40, 25. 6. 1984

3.43   Brinkmeyer F M et al., Process boosts 95% selectivity for LPG, Oil & Gas J. 81, p 75, 28. 3. 1983

3.44   Pujado P R, Vora B V (1983) Chem. Economy & Eng. Rev. 15, No 5: 16

3.45   Aguilo A, Hobbs C C, Zey E G (1985) Acetic acid, Ullmann (5), vol 1, p 50, Verlag Chemie, Weinheim

3.46   Schwerdtel W (1970) Hydrocarbon Proc. 49, No 11: 117

3.47   De Maio D A, Will butane replace benzene as a feedstock for maleic anhydride?, Chem. Eng. 87, p 104, 19. 5. 1980

3.48   Malow M (1980) Hydrocarbon Proc. 59, No 11: 149

3.49   Budi F et al. (1982) Hydrocarbon Proc. 61, No 1: 159

3.50   Arnold S C et al. (1985) Hydrocarbon Proc. 64, No 9: 123

3.51   Alma Process, Petrochemical Process Handbook (1985) Hydrocarbon Proc. 64, No 11: 142

3.52   New route to maleic anhydride, Chem. Week, p 24, 26. 8. 1987

3.53   Di Cio A, Verde L (1985) Hydrocarbon Proc. 64, No. 8: 68

3.54   Butane beating benzene as maleic feed, Chem. Eng. 88, p 25, 2. 11. 1981

3.55   Oxirane's C4 chemistry is key to European expansions, Europ. Chem. News 37, p 23, 5. 3. 1979

3.56   Hasuike T, Matsuzawa H (1979) Hydrocarbon Proc. 58, No. 2: 105

3.57   Porcelli R V, Juran B (1986) Hydrocarbon Proc. 65, No 3: 37

3.58   Seeboth H et al. (1978) Chem. Tech. 30: 575

3.59   Seeboth H et al. (1978) Chem. Tech. 30: 465

3.60   Weitkamp J, Edye E (1980) Erdoel u. Kohle 33: 16

3.61   Weitkamp J, Maixner S (1983) Erdoel u. Kohle 36: 523

3.62   Pass F (1981) Verarbeitung von Erdoel, Winnacker-Kuechler (4), vol 5, p 106, Hanser, Muenchen

3.63   Meyer D W, Chapin L E, Muir R T (1983) Chem. Eng. Prog. 79, No. 8: 59

3.64   Schulze J, Weiser M (1985) Vermeidungs- und Verwertungsmoeglichkeiten von Rueckstaenden bei der Herstellung chlororganischer Produkte, Research report no 103 01 304, Texte 5/85, p 201, Umweltbundesamt, Berlin

3.65   Johnson J A, Mowry J R, Anderson R F (1985) The Cyclar Process, UOP Handbook of Technology Conference, p D-5, London

3.66   Philpot J A, The eternal problem: What to do with C4's?, Europ. Chem. News 33, p 15, 17. 10. 1975

3.67   EPA gasoline rules yield a butane bonanza, Chem. Week, p 6, 12. 8. 1987

3.68   Felten J R, McCarthy K M (1987) Hydrocarbon Proc. 66, No 5: 49

3.69   Hamann, W (1982) Erdoel u. Kohle 35: 342

3.70   Franck, H-G, Stadelhofer J W (1987) Industrielle Aromatenchemie, Springer, Berlin Heidelberg New York

3.71   Guerico V J (1982) Erdoel u. Kohle 35: 253

3.72   Rossenbeck M et al. (1979) Motorkraftstoffe, Ullmann (4), vol 17, p 61, Verlag Chemie, Weinheim

3.73   Selectopol, Refining Process Handbook (1980), Hydrocarbon Proc. 59, No 9: 121

3.74   Schaefer W (1976) Der Lichtbogen 25, No 181: 4

3.75   Methyl tertiary butyl ether process can help separate all C4 fractions, Oil & Gas J. 77, p 76, 1. 1. 1979

3.76   Obenaus F, Droste W (1980) Erdoel u. Kohle 33: 271

3.77   The Snamprogetti/Anic MTBE Technology (1982) Chem. Economy & Eng. Rev. 14, No. 5: 37

3.78   Smith L A, Huddleston, M N (1982) Hydrocarbon Proc. 61, No 3: 121

3.79   Information from Butzert H, Huels, Marl, Feb. 1987

3.80   Fluor wins Arco's MTBE plant contract at Botlek, Europ. Chem. News 45, p 24, 20. 4. 1987

3.81   MTBE (1986) Informations Chimie, No 275: 187

3.82   Rossenbeck M et al. (1979) Motorkraftstoffe, Ullmann (4), vol 17, p 51, Verlag Chemie, Weinheim

3.83   Etherol, Refining Process Handbook (1986) Hydrocarbon Proc. 65, No 9: 105

3.84   Kroenig W, Scharfe G (1966) Erdoel u. Kohle 19: 497

3.85   Scharfe G (1973) Hydrocarbon Proc. 52, No 4: 171

3.86   Boucher J F et al., Dimersol X process makes octenes for plasticizer, Oil & Gas J. 80, p 84, 29. 3. 1982

3.87   Propylene: While worldwide demand grows, Chem. Week, p 26, 22. 7. 1987

3.88   Schulze J, Homann M (1986) Untersuchung der Vermeidungs- und Verwertungsmoeglichkeiten von schwefelhaltigen Rueckstaenden der anorganischen und organischen Zwischenproduktenchemie, Research report no 103 01 343, Texte 16/86, p 394, Umweltbundesamt, Berlin

3.89   Leonard J, Gaillard J F (1981) Hydrocarbon Proc. 60, No 3: 99

3.90   Dimersol X, Petrochemical Process Handbook (1985) Hydrocarbon Proc. 64, No 11: 15

3.91   Friedlander R G et al. (1986) Hydrocarbon Proc. 65, No 2: 31

3.92   Octenes, Petrochemical Process Handbook (1987) Hydrocarbon Proc. 66, No 11: 82

3.93   Weitz H M, Hartig J (1975) Butadien, Ullmann (4), vol 9, p 8, Verlag Chemie, Weinheim

3.94   Stobaugh R B (1967) Hydrocarbon Proc. 46, No 6: 141

3.95   Weitz H M, Hartig J (1975) Butadien, Ullmann (4), vol 9, p 9, Verlag Chemie, Weinheim

3.96   Welch L M, Croce L J, Christmann H F (1978) Hydrocarbon Proc. 57, No 11: 131

3.97   Husen P C, Deel K R, Peters W D, Phillip's butadiene process is success, Oil & Gas J. 69, p 60, 2. 8. 1971

3.98   Neier W, Strehlke G (1986) 2-Butanone, Ullmann (5), vol 4, p 478, Verlag Chemie, Weinheim

3.99   Smidt J, Krekeler H (1963) Erdoel u. Kohle 16: 560

3.100   Neier W, Strehlke G (1986) 2-Butanone, Ullmann (5), vol 4, p 475, Verlag Chemie, Weinheim

3.101   Broich F (1962) Chem.-Ing.-Tech. 34: 45

3.102   MA, Petrochemical Process Handbook (1971) Hydrocarbon Proc. 50, No 11: 175

3.103   Lenz G (1976) Herstellung von Maleinsaeureanhydrid aus Butenen, Chem. Anl. Verf., No 7, p 27

3.104   Writers St (1972) Chem. Economy & Eng. Rev. 4, No 4: 22

3.105   Brockhaus R (1971) Erdoel u. Kohle 24: 397

3.106   Hatch L F, Matar S (1978) Hydrocarbon Proc. 57, No 8: 153

3.107 Graefje H et al. (1986) Butanediols, Butenediol, and Butynediol, Ullmann (5), vol 4, p 460,
       Verlag Chemie, Weinheim
3.108 Gueterbock H (1959) Polyisobutylen und Isobutylen-Mischpolymerisate, Springer, Berlin
       Goettingen Heidelberg
3.109 Oelchemie 1984/1985, p 73, Deutsche Exxon Chemie, Koeln 1985
3.110 Pressindustria in Soviet butyl rubber venture, Europ. Chem. News 46, p 19, 4. 11. 1. 1988
3.111 BASF start gives MMA makers pause, Chem. Week, p 64, 20. 1. 1988
3.112 Dettmeier U et al. (1982) Aliphatische Zwischenprodukte, Winnacker-Kuechler (4), vol 6,
       p 118, Hanser, Muenchen
3.113 Acrylonitrile: How, where who − future (1971), Hydrocarbon Proc. 50, No 1: 109
3.114 Wenzel F, Lehmann K (1978) Methacrylsaeure und Methacrylate, Ullmann (4), vol 16, p 609,
       Verlag Chemie, Weinheim
3.115 New MMA-production process (1983) Chem. Economy & Eng. Rev. 15, No 1/2: 42
3.116 Nakamura T, Kita T (1983) Chem. Economy & Eng. Rev. 15, No 10: 23
3.117 Methyl methacrylate, Petrochemical Process Handbook (1987) Hydrocarbon Proc. 66, No.
       11: 80
3.118 MMA (1983) Chem. Economy & Eng. Rev. 15, No 1/2: 36
3.119 Methacrylic acid, Petrochemical Process Handbook (1987) Hydrocarbon Proc. 66, No 11: 78
3.120 Big changes in store for methyl methacrylate, Chem. Eng. 85, p 25, 3. 7. 1978
3.121 Arco's neat fit: MMA and C4s, Chem. Week, p 41, 26. 8. 1987
3.122 Fell B, Heilen G (1976) Chem.-Ing.-Tech. 48: 485
3.123 Marwede G et al. (1982) Elastomere, Winnacker-Kuechler (4), vol 2, p 547, Hanser,
       Muenchen
3.124 Reis T (1972) Chem. & Proc. Eng. 53: 68
3.125 Swodenk W et al. (1970) Erdoel u. Kohle 23: 641
3.126 Giraud A L Y (1961) Chem. Eng. Prog. 57, No 9: 66
3.127 Naito K, Ogino K (1973) Chem. Economy & Eng. Rev. 5, No 1: 42
3.128 Disproportionation: A promising route to isoprene, Oil & Gas J. 69, p 172, 19. 4. 1971
3.129 Peterson H J, Turner J O (1974) Hydrocarbon Proc. 53, No 7: 121
3.130 Reis, T (1972) Chem. & Proc. Eng. 53: 66
3.131 Heath A, High-purity isoprene from acetone acetylene, Chem. Eng. 80, p 48, 1. 10. 1973
3.132 Polybutene-1, Oil & Gas J. 68, p 62, 23. 11. 1970
3.133 Kuhnen L (1971) Der Lichtbogen 20, No 161: 14
3.134 de Leeuw P W et al. (1980) Chem. Eng. Prog. 76, No 1: 57
3.135 Neste und Idemitsu entwickeln Polybuten-Verfahren (1985), Europa Chemie, No 34, p 589
3.136 Karol F J (1983) The polyethylene revolution, Chemtech 13, No 4, p 222
3.137 Wilson T M B (1985) Chem. Eng. Prog. 81, No 7: 16
3.138 LLDPE-Verfahren der Ruhrchemie (1983) Europa Chemie, No 13, p 218
3.139 Lineares Polyethylen geringer Dichte nach dem Du Pont-Verfahren (1984) Europa Chemie,
       No 7, p 104
3.140 Information from Doerrscheidt W, Huels, Marl, May 1987
3.141 ECN Chemscope International Project Reviev
3.142 IISRP (1984/1985) Rubber Statistical Bulletin 39, No 12, p 45
3.143 Hoffmann W (1986) Kunststoffe 76: 1150
3.144 Gorke K et al. (1970) Der Lichtbogen 19, No 155: 4
3.145 Bellringer F J, Hollis C E (1968) Hydrocarbon Proc. 47, No 11: 127
3.146 Fricke H (1980) Zwangsanfall von Rueckständen und ihre Beseitigungsmoeglichkeiten an
       Land am Beispiel Chlorkohlenwasserstoff-Rueckstaende, Muell Abfall 17: Extra issue, p 27
3.147 Nakatani H, Kato S (1972) Chem. Economy & Eng. Rev. 4, No 7: 30
3.148 Prescott J H, Butadiene to neoprene process makes U. S. debut, Chem. Eng. 78, p 47, 8. 2. 1971
3.149 Johnson P R (1979) Chloroprene, Kirk-Othmer (3), vol 5, p 781, Wiley, New York Chichester
       Brisbane Toronto
3.150 Rohde E (1982) Kautschuk u. Gummie Kunststoffe 35: 947
3.151 Lindenschmidt G, Theyson R (1987) Kunststoffe 77: 982
3.152 Weissermel K, Arpe H-J (1978) Industrielle Organische Chemie (2), p 117, Verlag Chemie,
       Weinheim
3.153 Weitz H M, Hartig J (1975) Butadien, Ullmann (4) vol 9, p 4, Verlag Chemie, Weinheim

3.154   Weber H (1977) Der Lichtbogen 26, No 186: 16
3.155   Davis, D D (1985) Adipic Acid, Ullmann (5), vol 1, p 272, Verlag Chemie, Weinheim
3.156   Weitz H M, Hartig J (1975) Butadien, Ullmann (4), vol 9, p 3, Verlag Chemie, Weinheim
3.157   Dettmeier U et al. (1982) Aliphatische Zwischenprodukte, Winnacker-Kuechler (4), vol 6, p 109, Hanser, Muenchen
3.158   Mueller K A, Wichard G (1970) Der Lichtbogen 19, No 155: 6
3.159   Ono I, Kihara K (1967) Hydrocarbon Proc. 46, No 8: 147
3.160   Tanabe Y (1981) Hydrocarbon Proc. 60, No 9: 187
3.161   Brownstein A M, List H L (1977) Hydrocarbon Proc. 56, No 9: 159
3.162   Graefje H et al. Butanediols, Butenediol, and Butynediol, Ullmann (5), vol 4, p 459, Verlag Chemie, Weinheim
3.163   Wachstumsmaerkte in traditionsreichem Gebiet (1986) Chem. Ind. 38: 756
3.164   Davy & Sohio link in gas-to-thermoplastics process, Europ. Chem. News 44, p 20, 17. 11. 1986
3.165   Feinauer R (1984) Der Lichtbogen 33, No 203: 4
3.166   Griesbaum K, Swodenk W (1980) Erdoel u. Kohle 33: 34
3.167   Schelde Chemie unit scrapped, Europ. Chem. News 42, p 6, 24. 9. 1984
3.168   Blank H U et al. (1982) Aromatische Zwischenprodukte, Winnacker-Kuechler (4), vol 6, p 148, Hanser, Muenchen
3.169   Voetter H, Kosters W C C (1966) Erdoel u. Kohle 19: 267
3.170   Lindstrom M, Williams R (1983) Sulfolanes and sulfones, Kirk-Othmer (3), vol 21, p 963, Wiley, New York Chichester Brisbane Toronto
3.171   Weitz H M, Hartig J (1975) Butadien, Ullmann (4), vol 9, p 5, Verlag Chemie, Weinheim
3.172   Franck H G, Stadelhofer J W (1987) Industrielle Aromatenchemie, p 403, Springer, Berlin Heidelberg New York
3.173   Frank H-G, Collin G (1968) Steinkohlenteer, p 152, Springer, Berlin Heidelberg New York
3.174   Mueller H J, Loeser E (1986) Butadiene, Ullmann (5), vol 4, p 433, Verlag Chemie, Weinheim
3.175   Mueller H J, Loeser E (1986) Butadiene, Ullmann (5) vol 4, p 434, Verlag Chemie, Weinheim
3.176   Arco and Shell exploit metathesis chemistry (1986) Proc. Eng., No 1, p 17

Chapter 4

4.1   Homann D, Hohmeier F (1973), Der Umgang mit statistischen Daten ueber den Kunststoffmarkt, Kunststoffe 63, Part 1: No 1, p 57; Part 2: No 2, p 123; Part 3a: No 3, p 187; Part 3b: No 4, p 257; Part 4: No. 5, p 343
4.2   Obenaus F, Droste W, Neumeister J (1986) Butenes, Ullmann (5), vol 4, p 487, Verlag Chemie, Weinheim
4.3   Obenaus F, Droste W, Neumeister J (1986) Butenes, Ullmann (5), vol 4, p 491, Verlag Chemie, Weinheim
4.4   Oelchemie 1984/1985, Deutsche Exxon Chemie, Koeln 1985

Chapter 5

5.1   Koelbel H, Schulze J (1970) Der Absatz in der Chemischen Industrie, p 25, Springer, Berlin Heidelberg New York
5.2   Acrylnitril-Butadien-Styrol-Copolymerisate (1986) Kunststoffe 76: 851
5.3   Poth L G (1973) Marketing in Fallstudien p 151, Verlag Moderne Industrie, Muenchen
5.4   Kollat D T, Blackwell R D, Robeson J F (1972), Strategic Marketing, p 4, Holt Rinehart & Winston, New York
5.5   Information from Werner V, Mineraloelwirtschaftsverband, Hamburg, Jan. 1988
5.6   Amecke H D (1987) Chemiewirtschaft im Ueberblick — Produkte, Maerkte, Strukturen, Verlag Chemie, Weinheim
5.7   Homann D (1979) Gummi Asbest Kunststoffe 32: 896
5.8   Lindenschmidt G, Theyson R (1987) Kunststoffe 77: 982
5.9   Koelbel H, Schulze J (1970) Der Absatz in der Chemischen Industrie, p 318, Springer, Berlin Heidelberg New York
5.10   Koelbel H, Schulze J (1970) Der Absatz in der Chemischen Industrie, p 103, Springer, Berlin Heidelberg New York

5.11 Koelbel H, Schulze J (1970) Der Absatz in der Chemischen Industrie, p 168, Springer, Berlin Heidelberg New York

5.12 Koelbel H, Schulze J (1970) Der Absatz in der Chemischen Industrie, p 653, Springer, Berlin Heidelberg New York

5.13 Riebel P (1970) Kuppelproduktion und -kalkulation, Management Enzyklopaedie, vol 3, p 1251, Verlag Moderne Industrie, Muenchen

5.14 Mellerowicz K (1966) Kosten und Kostenrechnung, vol 2, Part 1: Allgemeine Fragen der Kostenrechnung und Betriebsabrechnung (4), p 48, de Gruyter, Berlin

5.15 Riebel P (1970) Kuppelproduktion und -kalkulation, Management Enzyklopaedie, vol 3, p 1243, Verlag Moderne Industrie, Muenchen

5.16 Heidrich F (1982) Untersuchung der branchenbezogenen Eignung der Riebelschen Einzelkosten- und Deckungsbeitragsrechnung als Dispositions- und Kontrollinstrument, Diploma work, Technische Universität Berlin

5.17 Ehrt R, Zeitschr. f. Betriebsw. 53: 1197

5.18 Tillmann K H (1954) Zeitschr. f. handelsw. Forschung 48: 156

5.19 Heyn, P. (1973) Verrechnungspreise in der Chemischen Industrie, Zeitschr. f. betriebs. Forschung 25: Extra issue No 2, p 61

5.20 Koelbel H, Schulze J (1960) Projektierung und Vorkalkulation in der chemischen Industrie, p 368, Springer, Berlin Heidelberg New York

5.21 Schulze J (1987) Chem. Ind. 39, No 9: 36

5.22 Schulze J (1979) Chemie f. Labor u. Betrieb 30: 423

5.23 Kindler H, Nikles A (1979) Chem.-Ing.-Tech. 51: 1

5.24 Kindler H, Nikles A (1980) Kunststoffe 70: 802

Chapter 6

6.1   Winter F W (1975) Technische Waermelehre (9), p 177, Girardet, Essen

6.2   Erdoelinformationsdienst (1986) No 31, p 3

6.3   Little A D (1983) Auswirkungen des Wegfalls von Bleiverbindungszusaetzen und der Einstellung des Benzolgehalts von Benzin auf die deutschen Raffinerien, Part I, p 56, Umweltbundesamt, Berlin

6.4   Car registration statistics in the FRG (1987) Kraftfahrzeugbundesamt, Flensburg

6.5   Blumenstock K U (1986) Diesel kontra Benziner, mot-Technik, No 25, p 116

6.6   Heinrich U et al. (1986) J. of Applied Toxicology 6: 383

6.7   Dabelstein W, Reglitzky A (1986) Erdoel u. Kohle 102: 93

6.8   The rational use of energy in transportation (1982) CONCAWE, Den Haag

6.9   Little A D (1983) Auswirkungen des Wegfalls von Bleiverbindungszusaetzen und der Einstellung des Benzolgehalts von Benzin auf die deutschen Raffinerien, Part I and II, Umweltbundesamt, Berlin

6.10 Little A D (1983) Auswirkungen des Wegfalls von Bleiverbindungszusaetzen und der Einstellung des Benzolgehalts von Benzin auf die deutschen Raffinerien, Part I, p 70, Umweltbundesamt, Berlin

6.11 Little A D (1983) Auswirkungen des Wegfalls von Bleiverbindungszusaetzen und der Einstellung des Benzolgehalts von Benzin auf die deutschen Raffinerien, Part II, p 97, Umweltbundesamt, Berlin

6.12 Little A D (1983) Auswirkungen des Wegfalls von Bleiverbindungszusaetzen und der Einstellung des Benzolgehalts von Benzin auf die deutschen Raffinerien, Part I, p 112, Umweltbundesamt, Berlin

6.13 Mineraloelwirtschaftsverband, Annual report 1986, p 47, Hamburg

6.14 Little A D (1983) Auswirkungen des Wegfalls von Bleiverbindungszusaetzen und der Einstellung des Benzolgehalts von Benzin auf die deutschen Raffinerien, Part II, p 131, Umweltbundesamt, Berlin

6.15 Statistisches Bundesamt, Fachserie 4, Reihe 3.1, Meldenr. 221310

6.16 Weitkamp J, Edye E (1980) Erdoel u. Kohle 33: 16

6.17 Pierce V E, Logwinuk A K (1985) Hydrocarbon Proc. 64, No 9: 25

6.18 Chapin L E, Liolios G C, Robertson T M (1985) Hydrocarbon Proc. 64, No 9: 67

6.19 Meyer D W, Chapin L E, Muir R T (1983) Chem. Eng. Prog. 79, No 8: 59

6.20   Butamer, Refining Process Handbook (1980) Hydrocarbon Proc. 59, No 9: 168
6.21   Selectopol, Refining Process Handbook (1980) Hydrocarbon Proc. 59, No 9: 121
6.22   Methyl *tert* butyl ether, Refining Process Handbook (1980) Hydrocarbon Proc. 59, No 9: 215
6.23   Smith L A, Parker K E, MTBE aids efficiency of refinery C4's, Oil & Gas J. 78, p 124, 11. 8. 1980
6.24   Smith L A, Huddleston M N (1982) Hydrocarbon Proc. 61, No 3: 121
6.25   Pujado P R, Vora B V (1983) Chem. Economy & Eng. Rev. 15, No 5: 16
6.26   Arco, Annual report 1985
6.27   Horndasch G, Mohyuddin B I, Schliephake K (1985) Chem. Ind. 37: 780
6.28   O'Sullivan D A, Saudi chemicals buildup worries West European firms, Chem. & Eng. News, p 16, 4. 2. 1985
6.29   C4 isomerisation, Refining Process Handbook (1980), Hydrocarbon Proc. 59, No 9: 169
6.30   Craig R G, White E A (1980) Hydrocarbon Proc. 59, No 12: 111
6.31   Vora B V, Berg R C, Pujado P R (1983) Chem. Economy & Eng. Rev. 15, No 4: 27
6.32   *sec*-Butanol, Petrochemical Process Handbook (1987), Hydrocarbon Proc. 66, No 11: 67
6.33   Armstrong A P (1978) The octane requirement of European and Japanese cars, SAE Paper No. 780425, Warrendale, PA
6.34   Oxinol 50, Technical Bulletin, Arco Chemical Europe Inc. 1985
6.35   Pecci G, Floris T (1977) Hydrocarbon Proc. 56, No 12: 98
6.36   Oxygenates (1986) Brochure of the European Fuel Oxigenates Association (EFOA), Brussels
6.37   Little A D (1983) Auswirkungen des Wegfalls von Bleiverbindungszusaetzen und der Einstellung des Benzolgehalts von Benzin auf die deutschen Raffinerien, Part I, p 86, Umweltbundesamt, Berlin
6.38   Ludlow B, Methyl tertiary butyl ether production grows dramatically, Oil & Gas J. 85: 54, 8. 6. 1987
6.39   Rohe D (1986) Chem. Ind. 38: 570
6.40   Dabelstein W, Reglitzky A (1986) Erdoel u. Kohle 102: 246
6.41   Herwig J, Schleppinghoff B, Schulwitz S (1984) Hydrocarbon Proc. 63, No 6: 86
6.42   Fina, Total bring on line commercial TAME plant in U.K., Oil & Gas J. 85: 64, 19. 10. 1987
6.43   Chase J D, Galvez B B (1981) Hydrocarbon Proc. 60, No 3: 89
6.44   Veretherung von Crackbenzin (1986) Erdoel u. Kohle 39: 254
6.45   Schulze J, Homann M (1986) Untersuchung der Vermeidungs- und Verwertungsmoeglichkeiten von schwefelhaltigen Rueckstaenden der anorganischen und organischen Zwischenproduktenchemie, Research report no 103 01 343, Texte 16/86, p 274, Umweltbundesamt, Berlin
6.46   Kroeper H, Schloemer K, Weitz H M (1969) Erdoel u. Kohle 22: 605
6.47   Hoefermann H et al. (1970) Chem. Ind. 22: 32
6.48   Adler M S, Johnson D R (1979) Chem. Eng. Prog. 75, No 1: 77
6.49   OlefinSiv, Refining Process Handbook (1980) Hydrocarbon Proc. 59, No 9: 217
6.50   Strigle R E, Fukuyo K (1986) Hydrocarbon Proc. 65, No 6: 47
6.51   Butene-1, Petrochemical Process Handbook (1987) Hydrocarbon Proc. 66, No 11: 68
6.52   Sorbutene, Refining Process Handbook (1980) Hydrocarbon Proc. 59, No 9: 220
6.53   Wagner U, Weitz H M (1970) Ind. & Eng. Chem. 62, No 4: 43
6.54   Takao S (1966) Hydrocarbon Proc. 45, No 11: 131
6.55   Miller R L, Butadiene's Technical Shift, Chem. Eng. 79, p 52, 24. 1. 1972
6.56   Di Cio A, Verde L (1985) Hydrocarbon Proc. 64, No 8: 68
6.57   Malow M (1980) Hydrocarbon Proc. 59, No 11: 149
6.58   Butane beating benzene as maleic feed, Chem. Eng. 88: 25, 2. 11. 1981
6.59   Arnold S C et al. (1985) Hydrocarbon Proc. 64, No 9: 123
6.60   Information from Kraemer H, Huels, Marl, Apr. 1987
6.61   Kunststoffe im Automobilbau — Ersatz von Metallen kuenftig noch lohnender, Handelsblatt, p 15, 20. 7. 1981
6.62   Bakshi A S, MMA process shows economic promise, Oil & Gas J. 83: 99, 25. 11. 1985
6.63   Porcelli R V, Juran B (1986) Hydrocarbon Proc. 65, No 3: 37
6.64   Mitsubishi Rayon, Annual Report 1985
6.65   Kuhnen L (1971) Der Lichtbogen 20, No 161: 14
6.66   Dietrich J (1970) Der Lichtbogen 19, No 156: 7

6.67   Information from Neundorf U, Huels, Marl, May 1987
6.68   de Leeuw P W et al. (1980) Chem. Eng. Prog. 76, No 1: 57
6.69   Polybutene-1, Oil & Gas J. 68: 62, 23. 11. 1970
6.70   Polyethylene, Petrochemical Process Handbook (1980) Hydrocarbon Proc. 64, No 11: 160
6.71   ECN Chemscope International Project Review, Mar. 1985
6.72   Kunststoffrohre fuer Fussbodenheizungen (1984) Fussbodenheizungs-Sanitaerrohre 3, No 4: 135
6.73   Kunststoffwochendienst, Hamich W (ed) Darmstadt, 27. 3. 1987
6.74   Saechtling H (1986) Kunststoff Taschenbuch (23), p 231, Hanser, Muenchen
6.75   Information from Doerrscheidt W, Huels, Marl, May 1987
6.76   Kramer E, Koppelmann J (1983) Kunststoffe 73: 714
6.77   Karol F J (1983) Chemtech 13: 222
6.78   Optimismus in der Kunststoffindustrie (1983) Europa Chemie, No 27: 468
6.79   LLDPE to grab slice of Market, Europ. Chem. News 43, p 15, 9. 12. 1985
6.80   Information from Pickstroer J, Huels, Marl, May 1987
6.81   Viinanen J, All three polyethylenes, ECMRA-Conference Berlin, 14.–16. 10. 1985
6.82   Saechtling H (1986) Kunststoff Taschenbuch (23), p 227, Hanser, Muenchen
6.83   Borks S (1984) Kunststoffe 74: 474
6.84   Lineares Polyethylen niedriger Dichte (1984) Swiss Plast. 6: 29
6.85   Saechtling H (1986) Kunststoff Taschenbuch (23), p 220, Hanser, Muenchen
6.86   Kalle baut Polypropylenfolien-Kapazitaet aus (1984) Europa Chemie, No 32: 544
6.87   ECN Plastics price report
6.88   Oelchemie 1984/1985, p. 119, Deutsche Exxon Chemie, Koeln 1985
6.89   Wilson T M B (1985) Chem. Eng. Prog. 81, No 7: 16
6.90   Oelchemie 1984/1985, p. 120, Deutsche Exxon Chemie, Koeln 1985
6.91   Saechtling H (1986) Kunststoff Taschenbuch (23), p 228, Hanser, Muenchen
6.92   Saechtling H (1986) Kunststoff Taschenbuch (23), p 227, Hanser, Muenchen
6.93   Lineares Polyethylen niedriger Dichte (1986) Kunststoffe 76: 840
6.94   Brenntag liefert Saudi-LLDPE (1985) Europa Chemie, No 7: 103
6.95   Information from Pfannmueller H, Rheinische Olefinwerke, Wesseling, Oct. 1984
6.96   Specialities seek out premium markets, Europ. Chem. News 43, p 16, 9. 12. 1985
6.97   UCC Makes VLDPE, Europ. Chem. News 42, p 23, 24. 9. 1984
6.98   Kashiwa, H, Toyota A (1986) Chem. Economy & Eng. Rev. 18, No 10: 14
6.99   Ho S Y, O'Donnel J H (1984) Europ. Polymer J. 20: 421

Company Index

Subject Index

H.-G. Franck,
J. W. Stadelhofer,
Frankfurt, FRG

Industrial Aromatic Chemistry

Raw Materials – Processes – Products

1988. 206 figures, 88 tables, 720 structural formulas. XIV, 486 pages. ISBN 3-540-18940-8

From the contents: The Nature of the Aromatic Character. – Base Materials for Aromatic Chemicals. – Production of Benzene, Toluene and Xylenes. – Production and Uses of Benzene Derivatives. – Production and Uses of Toluene Derivatives. – Production and Uses of Xylene Derivatives. – Polyalkylated Benzenes. – Naphthalene – Alkylnaphthalenes and Other Bicyclic Aromatics. – Anthracene. – Further Polynuclear Aromatics. – Production and Uses of Carbon Products from Mixtures of Condensed Aromatics. – Aromatic Heterocycles – Toxicology/Environmental Aspects. – The Future of Aromatic Chemistry.

Aromatic organic hydrocarbons and heterocycles represent a bulk of about one third of all industrially produced organic basic materials. Aromatic compounds such as benzene, phenol, naphthalene, anthracene, and their homologues, are derived from raw materials, coal, crude oil and biogenic resources by thermal and catalytic refining processes.
This book introduces the chemistry of aromatics with a brief discussion of the aromatic character and a survey of historical aspects, particularly the development of the organic dye industry during the 19th century. The main emphasis of the book is to give a clear prospect of industrial processes for the production and the derivatisation of aromatics with consistent flow diagrams. Economical aspects of by- and side-products are especially regarded. For the most important aromatics an analysis of the international market included their derivatives: polymers, pesticides, dyes, pigments and drugs.
Professional scientists, managers and students in chemistry and chemical engineering will find a wealth of information for their career and daily work.
The book is an adapted and updated translation of the German edition.

Springer-Verlag Berlin
Heidelberg New York London
Paris Tokyo Hong Kong

Deutsche Ausgabe:

H.-G. Franck, J. W. Stadelhofer: **Industrielle Aromatenchemie.** 1987. ISBN 3-540-18146-6

M. B. Hocking,
University of Victoria, B. C.,
Canada

Modern Chemical Technology and Emission Control

1985. 152 figures. XVI, 460 pages. ISBN 3-540-13466-2

Contents: Background and Technical Aspects of the Chemical Industry. – Air Quality and Emission Control. – Water Quality and Emission Control. – Natural and Derived Sodium and Potassium Salts. – Industrial Bases by Chemical Routes. – Electrolytic Sodium Hydroxide and Chlorine and Related Commodities. – Sulfur and Sulfuric Acid. – Phosphorus and Phosphoric Acid. – Ammonia, Nitric Acid and their Derivatives. – Aluminium and Compounds. – Ore Enrichment and Smelting of Copper. – Production of Iron and Steel. – Production of Pulp and Paper. – Fermentation Processes. – Petroleum Production and Transport. – Petroleum Refining. – Formulae and Conversion Factors. – Subject Index.

F. Asinger,
Aachen, BRD

Methanol – Chemie- und Energierohstoff
Die Mobilisation der Kohle

1986. 93 Abbildungen, X, 407 Seiten.
ISBN 3-540-15864-2

Inhaltsübersicht: Methanol – Chemie- und Energierohstoff. – Herstellung von Synthesegas. – Die Methanolsynthese. – Methanol als alternativer Kraftstoff. – Das Energiemethanol. – Die Überführung von Methanol in Gemische von Paraffinkohlenwasserstoffen, Olefinen und von Aromaten. – Die Überführung von Methanol in technische Gase. – Andere Verfahren der Kohleveredlung als Alternativen zum Methanol. – Die industrielle Herstellung von organischen Chemikalien aus Methanol. – Eiweiß durch bakterielle Umsetzung von Methanol (SCP = Single-Cell-Protein). – Sachverzeichnis.

Springer-Verlag Berlin
Heidelberg New York London
Paris Tokyo Hong Kong

Vertriebsrechte für die sozialistischen Länder:
Akademie-Verlag Berlin